cSUR-UT Series: Library for Sustainable Urban Regeneration
Volume 1

Series Editor: Shinichiro Ohgaki, Tokyo, Japan

cSUR-UT Series: Library for Sustainable Urban Regeneration

By the process of urban development in the 20th century, characterized by suburban expansion and urban redevelopment, many huge and sophisticated complexes of urban structures have been erected in developed countries. However, with conventional technologies focused on the construction of structures, it has become difficult to keep urban spaces adaptable to environmental constraints and economic, social and cultural changes. In other words, it has become difficult for conventional technologies to meet social demands for the upgrading of social capital in a sustainable manner and for the regeneration of attractive urban space that is not only safe and highly efficient but also conscious of historical, cultural and local identities to guarantee a high quality of life for all. Therefore, what is needed now is the creation of a new discipline that is able to reorganize the existing social capital and the technologies to implement it.

For this purpose, there is a need to go beyond the boundaries of conventional technologies of construction and structural design and to integrate the following technologies:

(1) Technology concerned with environmental and risk management
(2) Technology of conservation and regeneration with due consideration to the local characteristics of existing structures including historical and cultural resources
(3) Technologies of communication, consensus building, plan making and space management to coordinate and integrate the individual activities initiated by various actors of society

Up to now, architecture, civil engineering, and urban engineering in their respective fields have, while dealing with different time-space scales and structures, accumulated cutting-edge knowledge and contributed to the formation of favorable urban spaces. In the past, when emphasis was put on developing new residential areas and constructing new structures, development and advancement of such specialized disciplines were found to be the most effective.

However, current problems confronting urban development can be highlighted by the fact that a set of optimum solutions drawn from the best practices of each discipline is not necessarily the best solution. This is especially true where there are relationships of trade-offs among such issues as human risk and environmental load. In this way, the integration of the above three disciplines is strongly called for.

In order to create new integrated knowledge for sustainable urban regeneration, the Center for Sustainable Urban Regeneration (cSUR), The University of Tokyo, was established in 2003 as a core organization of one of the 21st Century Centers of Excellence Programs funded by the Ministry of Education and Science, Japan, and cSUR has coordinated international research alliances and collaboratively engages with common issues of sustainable urban regeneration.

The cSUR series are edited and published to present the achievements of our collaborative research and new integrated approaches toward sustainable urban regeneration.

K. Hanaki (Ed.)

Urban Environmental
Management and Technology

🐎 Springer

Keisuke Hanaki, Ph.D.
Professor
Department of Urban Engineering,
The University of Tokyo
7-3-1 Hongo, Bunkyo-ku,
Tokyo 113-8656, Japan

Cover photo: Hangzhou, China; © Reiko Hayashi

ISSN 1865-8504
ISBN 978-4-431-78396-1   e-ISBN 978-4-431-78397-8

Library of Congress Control Number: 2008923161

Springer is a part of Springer Science+Business Media
springer.com
Printed in Japan

Typesetting: Camera-ready by the editors and authors
Printing and binding: Shinano Inc., Japan

Printed on acid-free paper

Preface

The environmental aspect of cities is nowadays well recognized as a critical element of urban development, management and regeneration. There are various environmental issues related to cities, and they have been analyzed in an individualized manner. Many types of technologies, such as wastewater treatment, have been developed to solve particular environmental problems. However, many of these problems are related to each other, and social and economic aspects are also important for urban regeneration.

A holistic view combining knowledge of various urban environmental factors such as water, heat, energy, air, materials and waste, and a practical approach based on such understandings are essential to manage contemporary urban environmental issues, from a local scale to a global scale.

The University of Tokyo has been conducting the 21st Century COE (Centers of Excellence) Program on Sustainable Urban Regeneration. Transdisciplinary courses have been offered for graduate students from the departments of Urban Engineering, Civil Engineering and Architecture to present a holistic view of urban regeneration.

This book is based on the contents of the transdisciplinary course on environmental management and technology. I hope that its content will be useful for undergraduate and graduate students and for experts and policy makers in developed and developing countries.

Shinichiro Ohgaki
Project Leader, COE Program on Sustainable Urban Regeneration
The University of Tokyo

Contents

Part III Heat and Energy Management

Part I Water Environment

1. Water Management in Sustainable Buildings

Motoyasu Kamata

Masayuki Mae

Department of Architecture, The University of Tokyo
7-3-1, Hongo, Bunkyo-ku, Tokyo 113-8656, JAPAN

1.1 Introduction

In Japan, since World War II, better water systems have been developed and the amount of water used has increased rapidly. Nationally, rainfall is relatively plentiful and the cost of water service is not expensive in many regions. So people are not very sensitive about water conservation in regards to buildings. In that context, the main interest of designing the water system in a building has been to assure comfort by determining the proper number of fixtures and the system capacity.

In some regions like northern Kyushu or Shikoku, however, rainfall is scarce and severe water shortages happen almost annually. In those areas, water saving technology is considered very important. Recently, water shortage has become a global issue, and water saving technology is considered very important.

In this chapter, a survey of water consumption in Japanese buildings will be presented. Then, water saving technologies will be introduced. Mainly there are two methods to reduce water consumption. The first is to develop a fixture and system which consumes less water. The second is to make efficient use of rain and grey (recycled) water. If a "sustainable building" is being designed, selecting the right piping material and system is also crucial, so some new methods for cold water supply and sewage will also be presented in this paper.

1.2 Water equipment in old Japanese houses

1.2.1 Toilets

Before WWII, the usual toilet in Japan was an outhouse, not a water closet. Excrement was stored in wooden or ceramic pots and used as fertilizer.

After WWII, as the water service and the sewerage spread, water closets became popular. At first, squat toilets (Fig.1.1) were the most popular partly because people hesitated to touch a seat others had sat on. But as time passed, sitting toilets (Fig. 1.2) became preferred. Lately, squat toilets are becoming the ones only adopted in the public buildings.

Fig.1.1. Traditional squat toilet   Fig.1.2. Sitting toilet

1.2.2 Baths

Before WWII, baths were installed only in luxurious houses. Most people used public baths. Private baths for ordinary houses were introduced in 1955, in "Jutaku Kodan" ("Public Corporation for Housing", now "Urban Renaissance") apartments. Baths were the first fixtures to use hot water in ordinary Japanese houses (Fig. 1.3).

If you wanted to use the bath in Fig. 1.3, you would have to fill the wooden bathtub with cold water first. Then you would light the gas-boiler to heat the water. The bath is extremely small, so it could fit in the very small space reserved for the bathroom. It is also very dangerous, however, because the built-in gas boiler inhales the indoor air for combustion and sometimes releases lethal CO gas into the bathroom. There is also no safety from over-heating. No shower is attached. By today's standards, this kind of boiler is extremely inconvenient and perilous.

At that time, however, these kinds of baths were welcomed with enthusiasm by middle-class families, and were considered the most important hot water appliance. In time, bathtubs became larger, and used newer materials like FRP (Fibre-Reinforced Plastic) (Fig. 1.4). Newer versions also have an attached shower.

Fig. 1.3 Bathtub from late 1955

Fig. 1.4 FRP Bathtub from 1970

1.2.3 Kitchens

In traditional houses, the kitchen was a place only for women and not considered an important space when compared with the rooms for the husband or guests. So the kitchen was placed in the northern part of the house, a rather dark and cold place (Fig. 1.5). Many houses did not have any water faucets, so women had to bring water buckets in by hand. After WWII, the liberation of women became very important, and modernizing the kitchen was the one of most interesting issues for architects. In "Jutaku Kodan" apartments, the kitchen was moved to the more comfortable, brighter southern part of the house and equipped with modern items like sanitary stainless steel sinks and electric rice cookers. To save space, the kitchen and dining room were merged into the "Dining Kitchen" (DK). By these improvements, the DK was expected to function as the centre of family members' daily life and became the symbol of modern housing.

Hot water appliances in the kitchen were introduced in the 1960s. A small kitchen gas boiler was developed and set up for washing dishes (Fig. 1.6). These boilers had the same problems the bath boilers had. Kitchen boilers consumed indoor air as they burned and emitted CO_2 and CO gases into the kitchen, which polluted the room environment. There was no control of hot water temperature, nor any safety measures to avoid imperfect combustion. Despite these faults, these gas boilers were welcomed by housewives and became a common sight.

Fig. 1.5 A kitchen in 1930

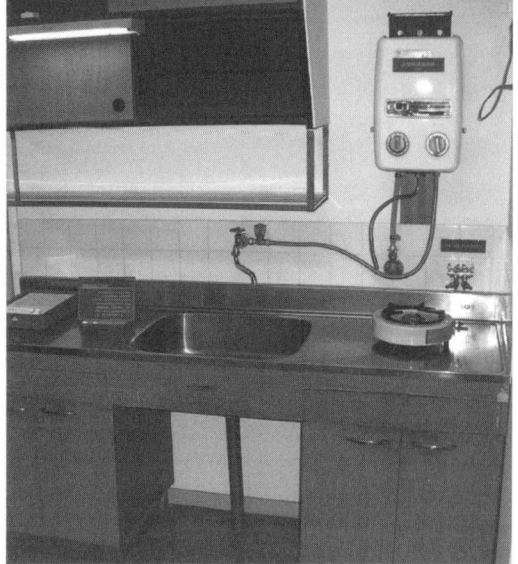

Fig. 1.6 A kitchen in 1960 with a gas boiler

1.3 Present Condition of Water Usage Japan

1.3.1 Cold Water Usage

As the water closet, bath and other appliances pervade, daily life has become fairly convenient for many people. But these improvements to the living standard also raised the volume of water usage in buildings.

The increase in the amount of "domestic water" is shown in Fig. 1.7. This value includes not only the amount used at home, but also the one consumed in public places like commercial buildings, hospitals, restaurant or so. Recently, the amount of "domestic water" used per person per day is estimated at 320~330 litres.

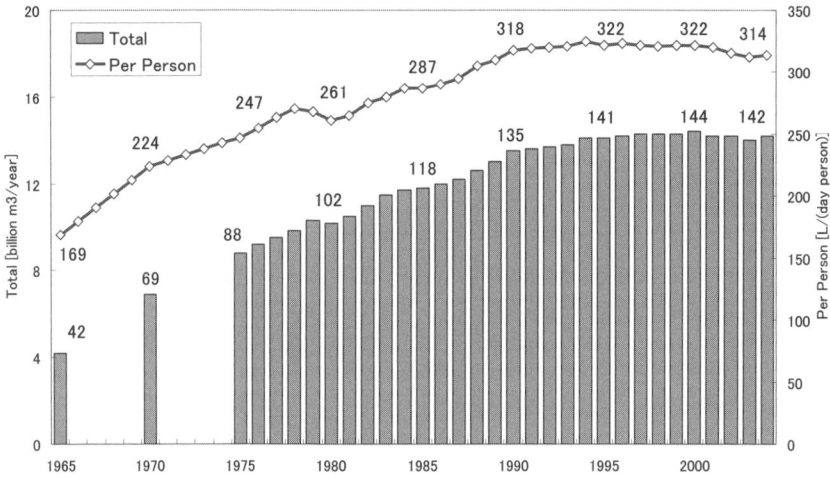

Fig. 1.7 Domestic water consumption (including housing and public buildings) [1]

Fig. 1.8 shows the water usage in public and residential buildings. Each amount is standardized by floor area (m^2) per day. It is obvious that the amount varies widely due to building type. The amounts in restaurants and cafés are prominent. The amount in terminal train stations is also huge, because many people gather and use the toilets. In houses or offices, the amount is relatively small.

Fig. 1.9 shows the breakdown of water usage according to type. In houses, the proportion of toilet, bath, laundry and kitchen are almost the same. In many building types, toilet water usage is the largest type. In department stores and restaurants, the kitchen consumes a considerable amount. In offices and hotels, the amount for air-conditioning is also high, which is mainly used in cooling towers for evaporation.

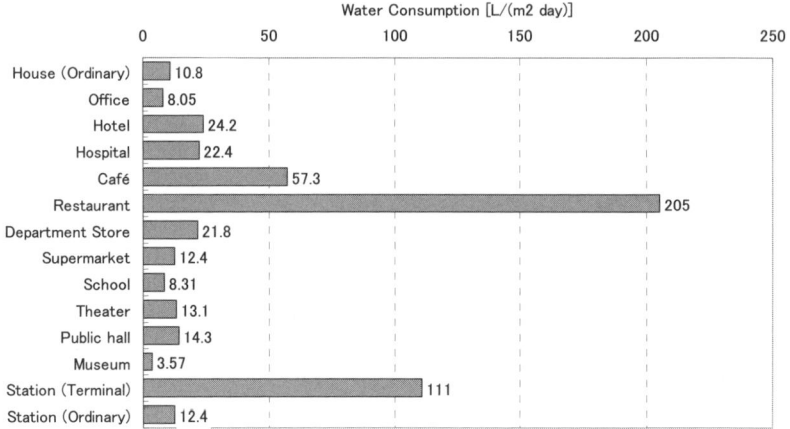

Fig. 1.8 Water consumption in public and residential buildings [2]

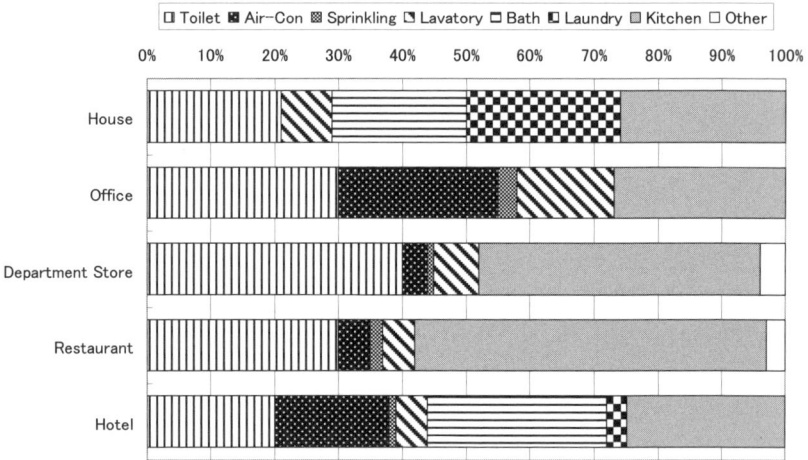

Fig. 1.9 Breakdown of water usage [3]

Fig. 1.10 shows the result of a recent survey by the committee in "Better Living" foundation regarding the amount of water used in one home per day, classified by house size. The total amount is 342.7 L/day in small 1K or 1DK homes, and 762.0 L/day in 3LDK homes. This shows that if the home is larger, the amount is larger. This is very understandable.

However, if you deal with the amount per person, a different trend appears. The amount of water used per person each day is also shown. More than 200L of water is used by each person daily, and this figure becomes larger in small houses often inhabited by just one person. In smaller houses, water in the bathtubs or laundry machines can be shared only by a small number of people. You can guess that that decreases utilization efficiency

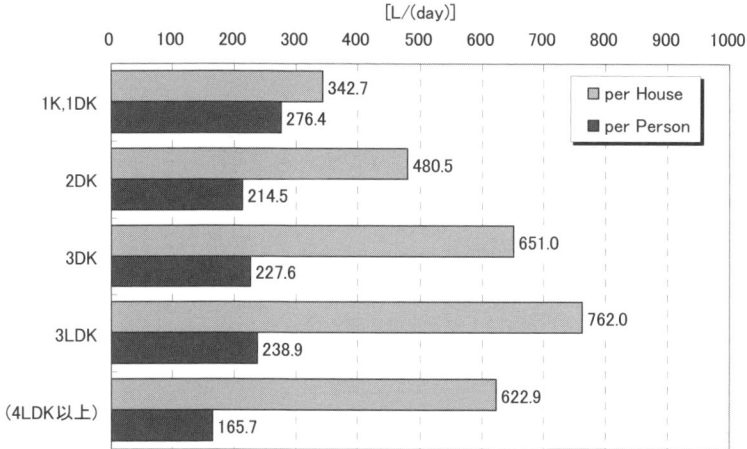

Fig. 1.10 Water consumption in residences (1 ~ 4: number of bedrooms, K: kitchen, DK: kitchen dining room, LDK: living-dining room with an attached kitchenette) [4]

1.3.2 Hot Water

As hot water equipment like bathtubs and kitchen boilers have gotten popular, hot water consumption has increased steadily. Hot water consumes not only water itself, but also a lot of energy like gas, electricity, or oil.

Fig. 1.11 shows the result of recent research on hot water consumption. Similar to cold water consumption, larger houses consume larger amounts. Houses with one person consume 186.8 L/day; houses with four people consume 444.9 L/day. However, if you measure the amount per person, 4-Person homes consume only 111.2 L/day, much smaller than 1-Person homes. This shows the same trend we already observed for cold water.

As you can guess, hot water consumption varies dramatically in different houses. Fig. 1.12 shows the volume and proportion of hot water consumption for each type of usage. The volume and breakdown of different types of water usage are widely different in each house.

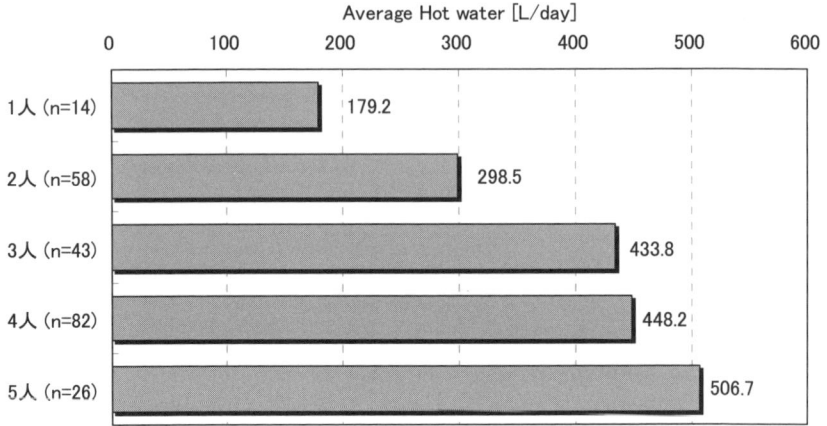

Fig. 1.11 Hot water consumption in residences [5]

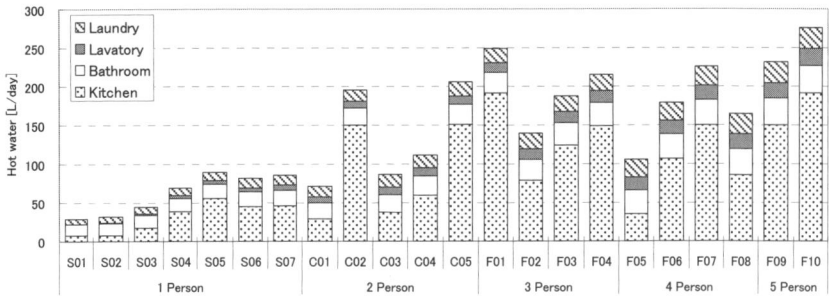

Fig. 1.12 Hot water consumption in houses (Annual Average) [6]

When the hot water supply system and equipment are designed, proper flow rate and temperature for each usage, like showers or kitchen faucets, are indispensable for safety and comfort.

For a long time, standard data was derived from American standards. But it was often criticized that Japanese preferences for hot water could be different from Americans'. So, a committee in The Society of Heating, Air-Conditioning and Sanitary Engineers of Japan (SHASE) carried out a test to determine the proper flow-rate and temperature for each usage. The result is shown in Table. 1.1. Now, these values are used widely to determine the capacity of boilers or diameter of pipes for buildings in Japan.

Table 1.1 Optimum temperature and flow rate for each type of water usage [7]

Usage		Optimum	
		Temperature [C^o]	Flow Rate [L/min]
Dish Washing	Ordinary faucet	39.0	7.8
	Shower type	39.0	5.0
Face washing		37.5	8.5
Laundry with hands		39.0	10.5
Shampoo		40.5	8.0
Taking a shower	hand-held	40.5	8.5
	attached to the wall	42.0	Fig 1.26

1.3.3 SHASE Standard

Until the 1970s, Japanese designers and engineers referred to the plumbing code of the USA for deciding the capacity of water system and sanitary condition in buildings. But as the difference of the actual state of Japanese buildings became obvious, SHASE (The Society of Heating, Air-Conditioning and Sanitary Engineers of Japan) established its own code based on the surveyed data of Japanese buildings.

The original plumbing code of Japan was called "HASS 206." This standard has been revised three times in 1982, 1991 and 2000. The name of HASS 206 was changed to "SHASE-S 206" in 2003. This manual is constantly referred to in designing the water system in buildings and became one of most important codes available for ensuring the sanitary condition of drinking water.

Fig. 1.13 HASS 206-2000

1.4 "CASBEE"(Comprehensive Assessment System for Building Environmental Efficiency)

The term "sustainable building" includes many aspects of architecture. Regarding water systems, it usually means efficient and environmentally-friendly systems which consume less energy.

In 1997, the Kyoto Protocol was announced to restrain the progress of global warming. According to that protocol, Japan has to decrease CO_2 emissions by 6 percent in 2010 compared to the rate of 1990. On the other hand, energy consumption in buildings has been increasing steadily and rapidly. To persuade building designers to consider saving energy, the "CASBEE" was developed [8]. CASBEE is the quantitative evaluation method to balance between building quality and environmental load. The former is called "Building Environmental Quality and Performance" (Q); the latter is called "Building Environmental Loadings" (L).

CASBEE can be used to evaluate the balance between "Q" and "L." If a building achieves a lower "Q" by emitting a higher "L," it is a low-performing "bad" building. If a building attains a higher "Q" through a lower "L," it is a high-performing "good" building. The balance of "Q" and "L" is called "BEE," and CASBEE's main objective is improving "BEE" to achieve good building performance by using less energy (Fig. 1.14).

The total scores are rated from "S," the best, to "A", "B" and "C," the worst. To achieve a higher score, designers are recommended to introduce "Reduction of Building Environmental Loadings" (LR) technologies.

Recently, the Japanese government has been very eager to reduce energy consumption in order to meet the goals of Kyoto Protocol. The Japanese Ministry of Land, Infrastructure and Transport strongly persuades developers to use CASBEE to measure energy performance and adopt more efficient and environmentally friendly systems and devices in buildings.

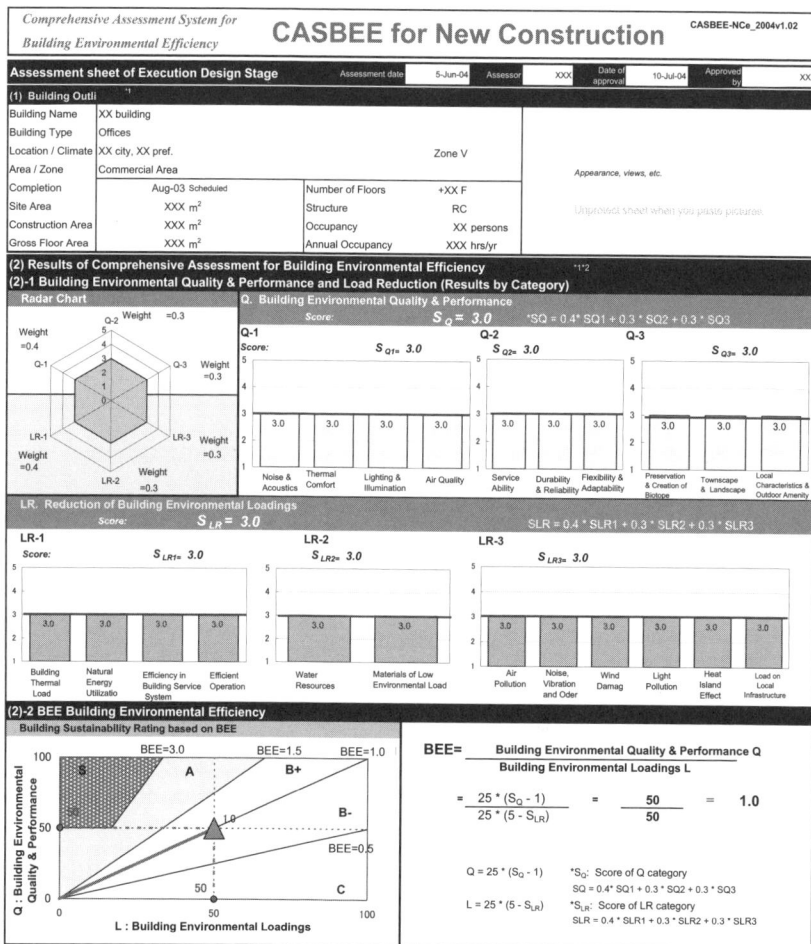

Fig. 1.14 CASBEE's total score

In the CASBEE system, water is an important factor. Fig. 1.15 shows the categories related to water systems in Q and LR. In Q-2, "Quality of Service," items like service life, reliability and flexibility are listed. In LR-1, "Energy," hot water efficiency is listed. In LR-2, "Resource & Materials," water saving technologies are listed.

(4) Score	Execution Design Stage				
		Entire Building and Common Properties		Residential and Accomodation sections	
Concerned categories		Score	weighting coefficients	Score	weighting coefficients	Total
Q Building Environmental Quality & Performance						3.0
Q-2 Quality of Service			0.30			3.0
2 Durability & Reliability			0.31		-	3.0
2.2 Service Life of Components			0.33		-	
3 Necessary Renewal Interval for Plumbing & Wiring Material			0.29		-	
2.3 Reliability			0.19		-	
2 Water Supply & Drainage			0.20		-	
3 Flexibility & Adaptability			0.29		-	3.0
3.3 Adaptability of Facilities			0.38		-	
2 Ease of Water Supply & Drain Pipe Renewal			0.17		-	
LR Reduction of Building Environmental Loadings						3.0
LR- Energy			0.40			3.0
3 Efficiency in Building Service System			0.30		-	3.0
3.4 Hot Water Supply System			0.05		-	
LR- Resources & Materials			0.30			3.0
1 Water Resources			0.15		-	3.0
1.1 Water Saving			0.40		-	
1.2 Rainwater & Gray Water			0.60		-	
1 Rainwater Use Systems			0.67		-	
2 Gray Water Reuse System			0.33		-	

Fig. 1.15 Water related Q and LR items

The details of each category in "Q" and "LR" are shown in Fig. 1.16, Fig. 1.17 and Fig. 1.18. If higher-level implements are selected, a higher score in each category can be obtained. CASBEE sums up each category's score and classifies the total score from "S" to "C."

Many items are evaluated in CASBEE, even only regarding water systems. Items covered range from pipe materials to water saving equipment and water recycling technology.

2.2.3 Necessary Renewal Interval for Plumbing & Wiring Materials	Weight (default)= 0.29
Level 3	Entire building and common properties
	Offices Schools Retailers Restaurants Halls Hospitals Hotels Apartments Factories
Level 1	(Inapplicable)
Level 2	(Inapplicable)
■Level 3	15 years
Level 4	16 years or more, less than 30 years
Level 5	30 years or more

Fig. 1.16 Service Life of Components (Durability & Reliability in Q-2)

2.3 Reliability

2.3.2 Water Supply & Drainage		Weight (default)=2.0
	Entire building and common properties	
	Offices Schools Halls Hospitals Hotels Apartments Factories	Retailers Restaurants
Level 1	None is applicable to the efforts to be evaluated.	None is applicable to the efforts to be evaluated.
Level 2	(Inapplicable)	(Inapplicable)
Level 3	Applicable to one of the efforts to be evaluated.	Applicable to one of the efforts to be evaluated.
Level 4	Applicable to two of the efforts to be evaluated.	(Inapplicable)
Level 5	Applicable to three or more of the efforts to be evaluated.	Applicable to two or more of the efforts to be evaluated

Efforts to improve the reliability of water supply & drainage		
Building Type	Offices Schools Halls Hospitals Hotels Apartments Factories	Retailers Restaurants
Score	Level 3	Level 3
	1) Water-saving equipment is used. This is limited to cases where it is used on a majority of the installed equipment. Water-saving devices are those approved as Eco Mark products, or	
	2) Plumbing systems are separated as far as possible to reduce the portions that become unserviceable in the event of a disaster.	
	3) The building has a pit for temporary waste water storage, in case main sewerage is unavailable after a disaster.	
	4) The building has two separate tanks, one for water reception and one elevated tank.	
	5) Planning enables the use of well water, rainwater, gray water etc.	
	6) Provision of a rainwater storage tank to provide domestic noncommercial water in the event of a disaster. . (Not applied to "Retailers" and "Restaurants.")	
	7) The building is equipped with a simple filtration system allowing conversion of rainwater to potable water in the event of a disaster. (Not applied to "Retailers" and "Restaurants.")	

Fig. 1.17 Reliability (Durability & Reliability in Q-2)

1   Water Resources

1.1   Water Saving   Weight(default)= 0.40

Level 3	Offices Schools Retailers Restaurants Halls Hospitals Hotels Apartments Factories
Level 1	No systems for saving water.
Level 2	(Inapplicable)
■Level 3	Major faucets are equipped with water-saving valve.
Level 4	In addition to water-saving valve, other water-saving equipment (such as flush-mimicking sound systems, water-saving toilets) is used.
Level 5	(Inapplicable)

1.2   Rainwater & Gray Water

1.2.1 Rainwater Use System   Weight(default)= 0.67

Level 3	Offices Schools Retailers Restaurants Halls Hospitals Hotels Apartments Factories
Level 1	(Inapplicable)
Level 2	(Inapplicable)
Level 3	No systems for using rainwater.
Level 4	Rainwater is used.
Level 5	Rainwater usage brings the rainwater usage rate to at least 20%.

1.2.2 Gray Water Reuse System   Weight(default)= 0.33

Level 3	Offices Schools Retailers Restaurants Halls Hospitals Hotels Apartments Factories
Level 1	(Inapplicable)
Level 2	(Inapplicable)
Level 3	No systems for reusing gray water.
Level 4	Gray water is reused.
Level 5	In addition to gray water reuse, there is equipment to reuse sewage.

Fig. 1.18 Water Resources (In LR-2)

1.5 Water saving technologies

The items adopted in CASBEE are considered very important components to realizing the goal of sustainable buildings. It is useful to know the details of the water saving technologies.

1.5.1 Toilets

Water saving closet and urinal

As already mentioned, toilets use a lot of water in all types of buildings, so the development of less water consuming toilets is the most important goal. But it is not very easy to implement because less water consumption means less cleansing and transportation ability. To achieve a good balance, the shape of the bowl and flushing methods are carefully designed (Fig. 1.19). By this effort, the water necessary to flush has decreased remarkably (Fig. 1.20). Adoption of these water saving toilets is very effective and strongly recommended. But the sewage system for the items, like pipe tilt, should be designed to assure good transportation performance.

Regarding urinals, the water flow pattern has been improved to ensure enough cleansing using less water (Fig. 1.21). If the cleansing is not thorough enough, clotting from urine can easily block the sewage pipes. So automatic flushing systems are designed carefully to flush minimal water periodically, even without usage.

Fig. 1.19 Water saving toilet

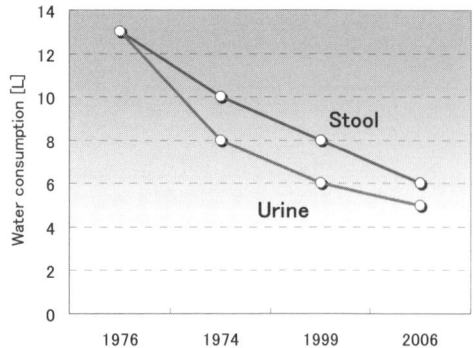

Fig. 1.20 Progression of water saving [9]

Sound mimicking device

In Japan, many women consume twice the volume of water than men when using the toilet. Because women want to conceal noise in the toilet, they flush the toilet soon after sitting. That means women flush the toilet at least twice. To avoid the waste, sound mimicking devices (Fig.1.22) have been introduced into women's stalls. This device makes a flushing sound, so there is no need to flush real water only to hide the noise.

Fig. 1.21 Water flow in urinal

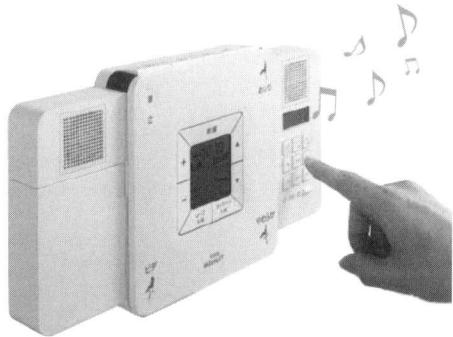

Fig. 1.22 Sound mimicking device

1.5.2 Residential Bathrooms

High-insulation bathtubs

As already mentioned, baths were the first hot water appliance in ordinary Japanese houses. Even these days, many Japanese prefer taking baths to showering.

To save energy consumption in the bathtub, improvements to heat insulation are effective. For Japanese, it is very common for family members to share the same hot water in the bathtub. But especially in winter, the hot water in the bathtub can get cold rapidly, so reheating is needed to keep a comfortable temperature.

High insulation bathtubs are designed to decrease the heat flow from inside by using double insulation on the bottom and side walls (Fig. 1.23). The top cover is also specially made to assure high performance. Compared with ordinary bathtubs, the temperature drop is decreased to less than one-third (Fig. 1.24).

Fig. 1.23 High-insulation bathtub

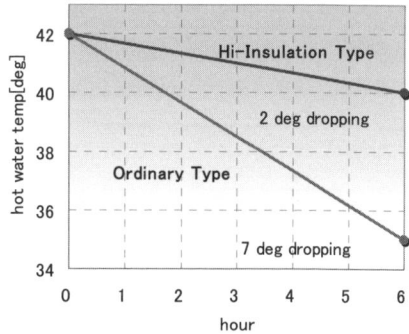

Fig. 1.24 Temperature drop [9]

Water-saving showerhead

Many people take a shower when bathing, and some young people prefer only showering now. To save hot water consumption in the shower, the development of a showerhead that satisfies users using a lower flow rate is important. Fig. 1.25 shows the relation between holes in the showerhead and optimum flow rate as the result of the examinees test. By adjusting the holes in the showerhead, optimum flow can be reduced without displeasure.

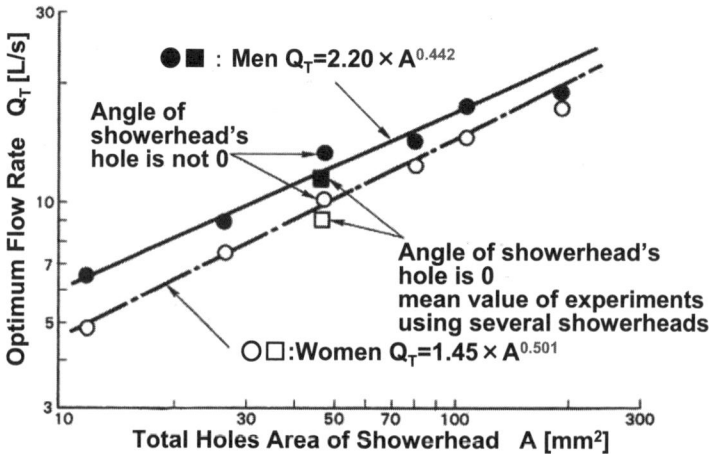

Fig. 1.25 Relation of total hole area and optimum flow rate of showerheads [10]

Some users are reluctant to pause the water during a shower even when they do not need it, because it is troublesome to turn the faucet's handle. Switch type showers (Fig. 1.26) were developed to eliminate that inconvenience. In some experiments, switch type showers reduced the water consumption by more than 20% in each season (Fig. 1.27).

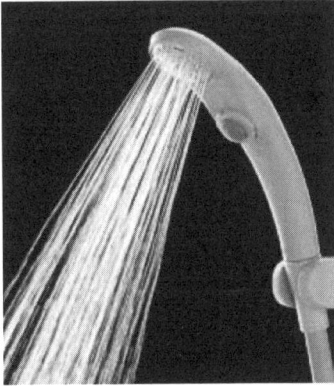

Fig. 1.26 Switch type shower

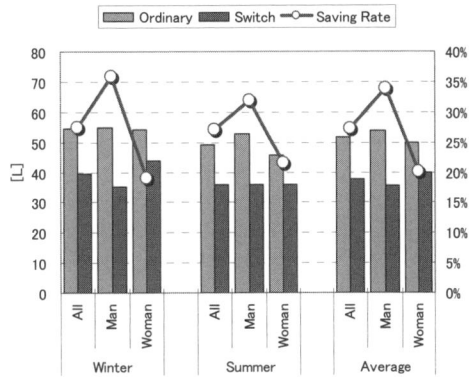

Fig. 1.27 Water saving effect [10]

1.5.3 Non-residential Lavatory

Automatic water saving faucet

Water saving in lavatories is important, especially for public buildings, which can be used by many guests. Automatic faucets are strongly recommended to avoid useless flowing. But in some cases, it is difficult to install because no electricity is available to power the automatic faucet. The faucet in Fig. 1.28 is a self power-generating faucet. Through water flow, power is generated and stored to operate itself (Fig. 1.29). This kind of faucet can be installed in any lavatory and is very effective in saving water.

Fig. 1.28 Self power-generating faucet

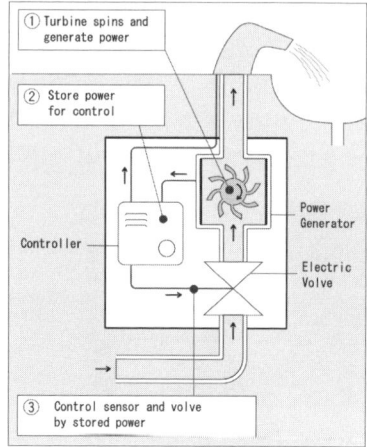

Fig. 1.29 Details of self power

1.5.4 Residential Kitchens

In houses, water consumption in the kitchen is considerable. Washing dishes uses the most. For housekeepers' convenience, some kinds of easy water stopping faucets have been developed. Some can stop water only by touching, with no need to turn the handle of the faucet (Fig. 1.30). Other devices can stop water by operating a switch with a hand or a foot (Fig. 1.31). It is very convenient to stop water by using your foot when your hands are busy washing dishes. Fig. 1.32 shows the result of an experiment using a water saving faucet. Both the regular touch and the foot-touch type can save water by 5 to 10 %.

Fig. 1.30 Easy touch type

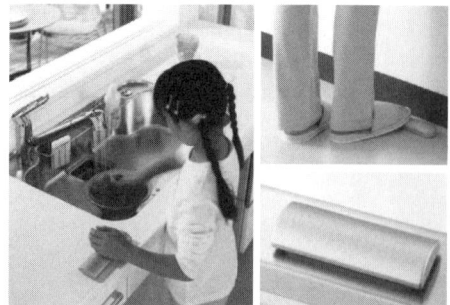

Fig. 1.31 Foot switch type

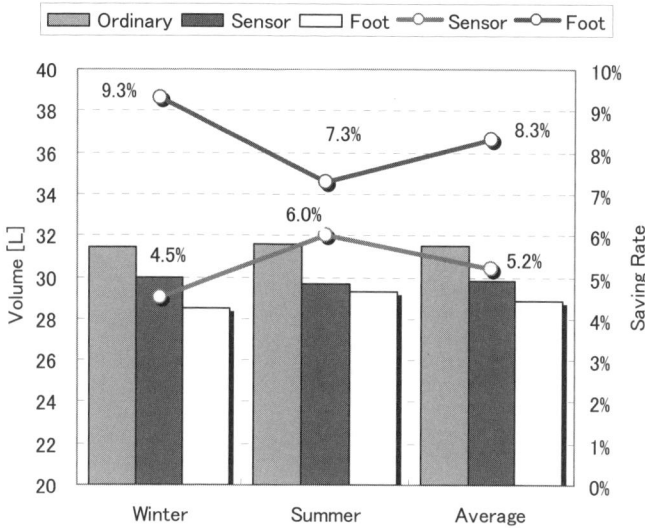

Fig. 1.32 Water saving effect of easy stopping kitchen faucet [11]

1.5.5 Residential Pipes

For piping in sustainable buildings, pipes need to be not only durable, but also replaceable. The structures of sustainable buildings are made strong enough to last for more than 100 years. So even the most durable piping must be replaced several times in the building's total life.

In old buildings, replaceable pipes were not a consideration, so it is a big burden to restore them. In the worst case, the building has to be totally destroyed. That illustrates the fact that initial planning is the very crucial to achieve a building with a long lifespan.

Cold and hot water supply

Since the 1970s, central water heating systems have become popular in Japanese houses. In old houses, small boilers were installed in the bath and kitchen for each type of usage, but the capacity of the boiler was restricted and boilers could cause indoor air pollution. As the demand for more comfort has risen, a central system with a high-capacity outdoor boiler has been introduced. A piping system connects the outdoor boiler and the faucets and hot water is available, even for the lavatory and laundry. A central system is very useful, but extended piping can cause problems. Early central systems used copper pipes that could corrode within several years, and it was very difficult to replace the old copper pipe.

Recently, header-conduit piping systems have been installed in many residences, especially in apartment buildings (Fig. 1.33 to 1.35). For that

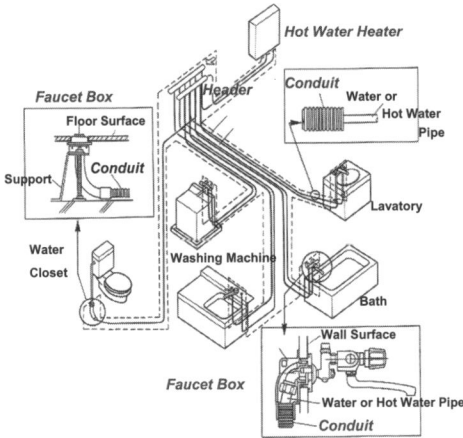

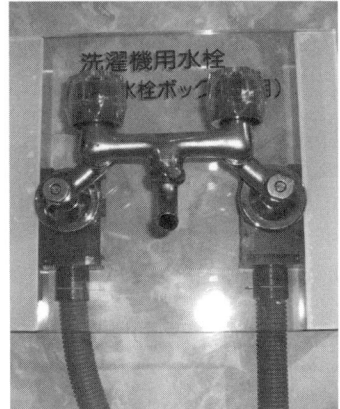

Fig. 1.34 Faucet

Fig. 1.33 Header-conduit piping system

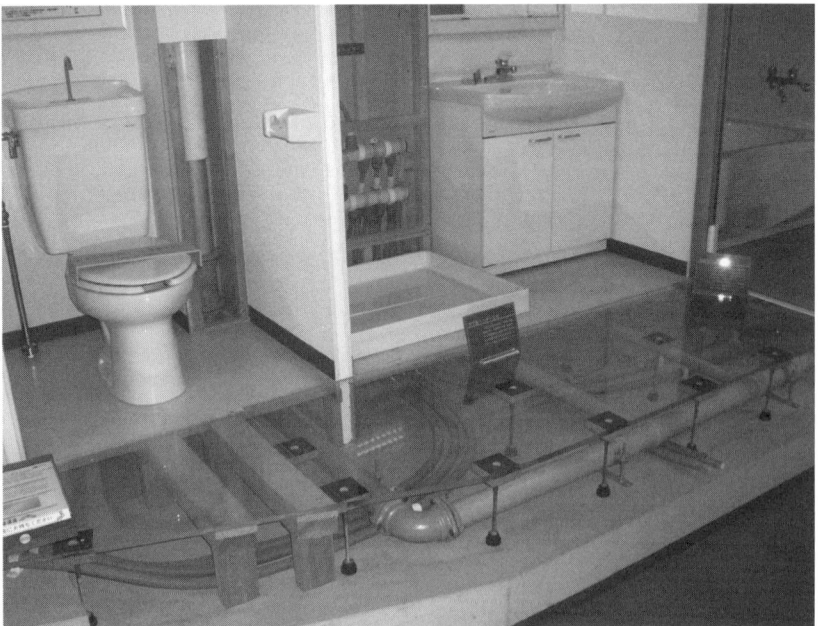

Fig. 1.35 Example of pipe system

system, the pipes are made of cross-linked polyethylene, a very durable material. The most important characteristic is that the pipe is covered by a conduit. Only the conduit is fixed in the floor or wall when constructed, so the inside pipe can be replaced through the conduit without damaging the floor or wall. With durable material and easy replaceability, this system assures a long serving life for the piping.

1.5.6 Sewage pipes

In old houses, sewage pipes, even the vertical ones, are installed in the central part of each house (Fig. 1.36). That means repair and replacement is very difficult.

To make replacement easier, vertical pipes have been moved outside of the common area. In this "Kodan Skelton Infill system" (KSI), developed by "Urban Renaissance," sewage piping is designed carefully using a header (Fig.1.38) and replaceable horizontal pipes (Fig.1.39) to assure a long life for the sewage system.

Fig. 1.36 Indoor Pipe in old houses   **Fig. 1.37** Outside sewage system of KSI

Fig. 1.38 Sewage header of KSI   **Fig. 1.39** "Easy to replace" layout

1.6 Water recycling technology in the supply system

Other than water saving fixtures and durable piping, water-recycling technologies are vital to create appropriate water systems for sustainable buildings. In this section, grey and rain water systems in real buildings are introduced.

1.6.1 Gray water treatment

NEXT21 (Fig. 1.40) is an experimental sustainable building, planned and operated by the Osaka Gas Corporation. The theme of this interesting apartment is the harmony of human life and the environment. For that goal, many technologies were developed to achieve a long life expectancy for the structure, equipment and so on.

The structure is composed of extremely strong columns and beams without a structural wall. That makes the renovation of the housing and equipment easy. To reduce the environmental impact, a ground garden (with pond) and roof garden were made and reserved for wild birds.

Fig. 1.40 Outlook of NEXT21

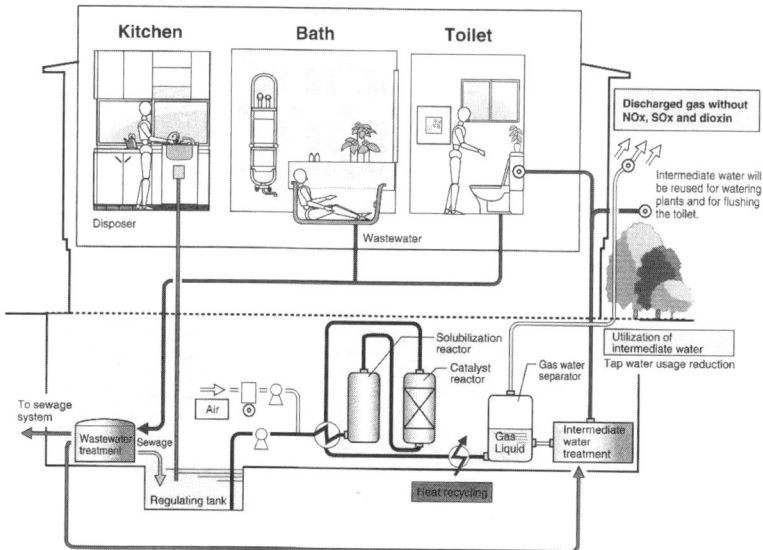

Fig. 1.41 "Aqua Loop" system in NEXT21

In NEXT21, a grey water treatment system called "Aqua Loop" has been adopted (Fig. 1.41). Sewage is collected and treated in the underground plant (Fig. 1.42). The purified water up to 8.5 m³/day nurtures the plants and flowers (Fig. 1.43). Other purified water of 3 m³/day is used for toilet flushing. This system can reduce water consumption and diminish the impact on the environment.

Fig. 1.42 Solubilization reactor   **Fig. 1.43** Pond in the garden

1.6.2 Use of Rainwater

A system that uses rainwater could be more efficient, because rainwater is much cleaner than sewage water. Also, a rainwater storing system in a building can reduce the outflow into the main sewage pipe during heavy rain that could cause flooding. In some areas afflicted with floods, this kind of system is recommended by the regional government.

Fig 1.44 shows the rainwater system in the "Ryogoku Kokugi-kan" building. This building collects rainwater by a large roof area (8,360 m²) and stores it in the 1,000 m³ underground tank. Treated rainwater makes up one-third of the total non-drinking water usage.

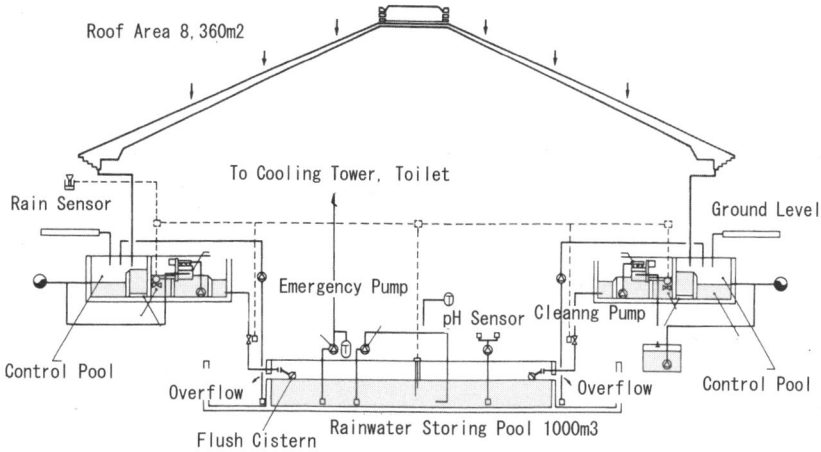

Fig. 1.44 Rainwater system in the "Ryogoku Kokugi-kan" building [3]

1.7 Conclusion

Because of the popularization of convenient water appliances in buildings, water consumption has been increasing constantly in Japan since WWII. On the other hand, the demand for building performance, including reduction of its environmental impact, has been growing as well. In that context, the design process of sustainable building now considers many aspects ranging from simple water reduction and recycling to piping material and replacement.

If buildings in regions with plenty of rainfall like Japan are the only ones considered, you may not think water related technology is very important compared with other types of energy usage like air conditioning. But in regions with scare rainfall, it is very essential. In light of the fact that scientists anticipate that global warming may cause water shortages in many regions worldwide, the importance of this kind of technology will be increasing in the near future.

References

[1] Ministry of Land, Infrastructure and Transport (2007), "Water resource in Japan"
[2] The Society of Heating Air-Conditioning and Sanitary Engineers of Japan (2001), "Practical knowledge of Sanitary Equipment 2nd Edition"
[3] The Society of Heating Air-Conditioning and Sanitary Engineers of Japan (1997), "Rain Water Utilization System"
[4] Based on the data collected in several apartments near Tokyo
[5] Masayuki MAE et al. (2007), "Survey on Actual Hot Water Consumption in Families of Various Sizes Part.1 Average Amount of Consumption in Each Family Size" Summaries of Technical Papers of Annual Meeting, Architectural Institute of Japan,
[6] Masayuki MAE et al. (2003), "THE ANALYSIS ON THE SEASONAL CHANGE IN USAGE OF HOT AND COLD WATER, The analysis and evaluation on usage of hot and cold water on urban apartment house (Part1)", JOURNAL OF ENVIRONMENTAL ENGINEERING (Transaction of AIJ) No.566, 2003.04
[7] Motoyasu KAMATA et al. (1993), "ABC of hot water", TOTO Publishing
[8] Japan Sustainable Building Consortium, "Comprehensive Assessment System for Building Environmental Efficiency CASBEE for New Construction - Technical Manual 2004 Edition", Institute for Building Environment and Energy Conservation (IBEC)
[9] Catalogue of TOTO Corporation, 2007
[10] Takeshi KONDO et al. (2006), "EFFECT OF THERMOSTATIC MIXING FAUCET AND SHOWER HEAD WITH STOP VALVE, Study on performance test of water saving fixture", JOURNAL OF ENVIRONMENTAL ENGINEERING (Transaction of AIJ) No.607, 2006.09
[11] Takeshi KONDO et al. (2007), "EFFECT OF WATER SAVING FIXTURE FOR DISHWASHING, Study on performance test of water saving fixture Part 2", JOURNAL OF ENVIRONMENTAL ENGINEERING (Transaction of AIJ) No.613, 2007.03

2. Urban Water Use and Multifunctional Sewerage Systems as Urban Infrastructure

Hiroaki Furumai

Department of Urban Engineering, The University of Tokyo
7-3-1, Hongo, Bunkyo-ku, Tokyo 113-8656, Japan

2.1 Introduction

2.1.1 Water Balance in Japan

Japan has a land area of 380,000 km^2 and a population of 120 million. An average precipitation of approximately 1,700mm/year provides about 650 billion m^3/year of water. The precipitation is much greater in Japan than the world average of 900mm/year. In Japan, one-third of the water disappears by evapotranspiration. Therefore, the available water resource is reduced to about 420 billion m^3/year. One-third of the water is runoff water from land to the sea during short periods, because of the rainy season and typhoons, leaving a stable water flow of one-third. This water balance is illustrated based on data from the fiscal year of 2000 in Fig.2.1. The artificial water flow is categorized into agriculture, industry and domestic water uses.

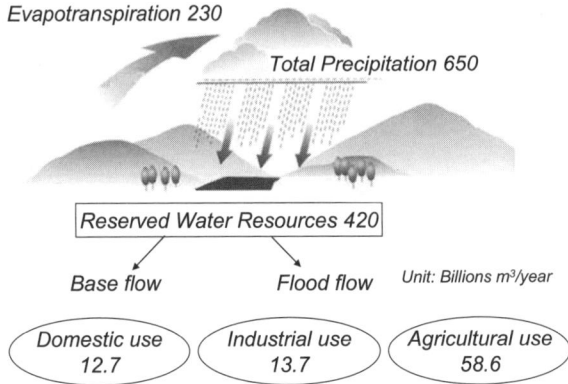

Evapotranspiration 230

Total Precipitation 650

Reserved Water Resources 420

Base flow   *Flood flow*   Unit: Billions m³/year

Domestic use   *Industrial use*   *Agricultural use*
12.7   *13.7*   *58.6*

Fig. 2.1. Water balance in Japan (data from FY 2000)

2.1.2 Urbanization and escalated waster use

Demographic changes have significant impacts on water use and its management. Fig.2.2 shows the past and predicted population growth in Japan and Southeast Asian countries from 1950 to 2025. It reveals a similarity in urban population growth from 1950 to 1970 in Japan and the coming future of Southeast Asian countries. This means that the early urbanization stage in Japan is being experienced in Southeast Asia; therefore, the history of water management in Japan could be useful to explore better ways of water management for Southeast Asia. A stable water supply and efficient water use have become concerns in mega cities. It is necessary to introduce new concepts, policies and measures for sustainable water use and management (Wagner et al.; 2002, Furumai, 2007).

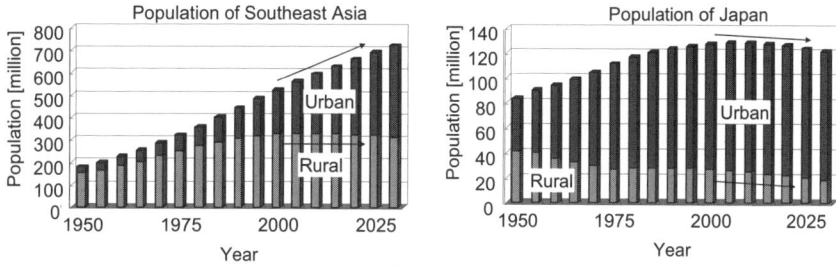

Fig. 2. 2. Urban and rural populations in Southeast Asia and Japan
Southeast Asia includes Cambodia, Indonesia, Lao People's Dem. Rep., Malaysia, Myanmar, Philippines, Thailand, and Vietnam.

Fig.2.3 shows representative locations with ground subsidence problems and their cumulative subsidence changes over time. In Tokyo, large scale groundwater pumping started in 1914. In the following years, the number of deep wells with a large diameter increased rapidly. Ground subsidence was first observed in Tokyo in the 1910s, and in Osaka in the 1920s.

The groundwater level continued to fall due to the extensive pumping to support increased production activities. Ground subsidence caused the destruction of buildings, damage from floods and high tides and rising public concern. By the mid-1940s, the damage to industry from World War II reduced the industrial use of groundwater, thereby halting ground subsidence. With the rapid progress of urbanization due to economic development after World War II, water supply had to catch up with remarkably increased water demand for domestic and industrial purposes. There was a sudden surge in demand for groundwater, and major subsidence took place again, especially in large urban areas.

Countermeasures against ground subsidence, e.g. control of groundwater pumping rates, were started in the 1960s and thereafter, the rate of subsidence in metropolitan areas slowed. Nowadays, measures have been introduced to control the abstraction of the water level and to avoid changes in the subsidence levels by optimizing water use and introducing new sources of water.

Urban water consumption has a major impact on the natural water cycle and consequently on the water environment. Therefore, understanding the urban water cycle, the potential of rainwater use and water recycling in urban areas is important for sustainability. In the case of mega cities, the sustainability of urban water use has become important in terms of a stable water supply and efficient water use. The achievement of sustainability is required to ensure a long-term water supply with adequate quality while minimizing adverse economic, social and environmental impacts.

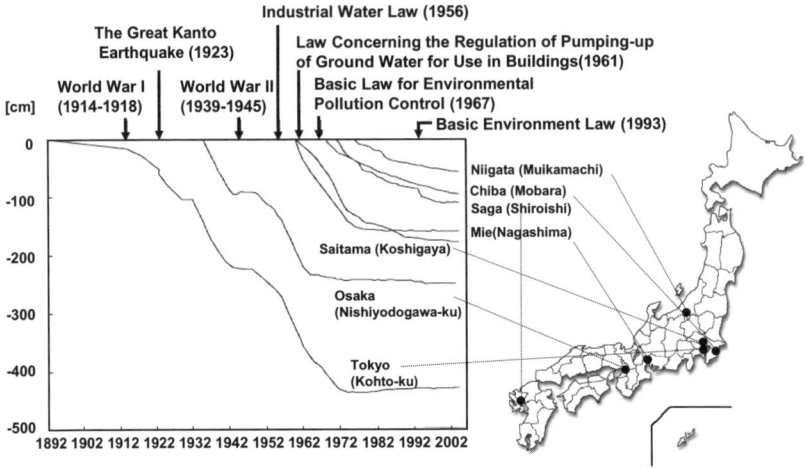

Fig. 2.3. Cumulative change in ground subsidence in typical areas over time

2.1.3 Water balance in Tokyo

Fig.2.4 shows the average temperature and rainfall in cities across the world: Tokyo, London, Cairo and Bangkok. Each city has different climate, weather, rainfall and population levels.

When the available water resource per capita from precipitation is calculated, Japan's value is much lower than the world average, or that of the USA, Australia or Switzerland, as shown in Fig.2.5. If the calculation is applied to the city and province levels, the available water resource supply in Tokyo and Kanagawa is at the same level as Saudi Arabia and Egypt.

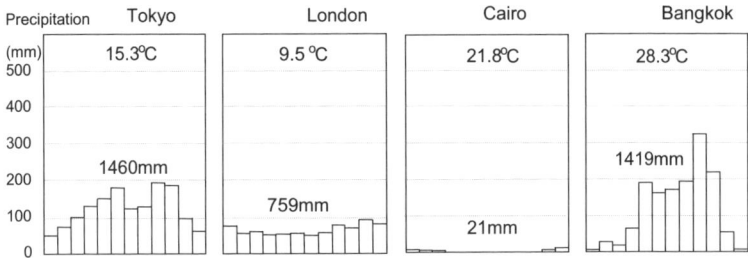

Fig. 2.4. Comparison of precipitation and temperature in several mega cities

m³/day/capita

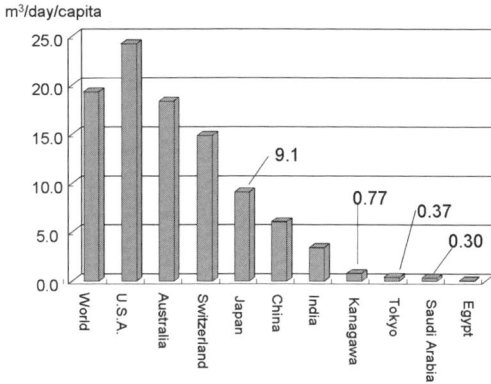

Fig. 2. 5. Comparison of available water resources per capita

Fig.2.6 presents the water balance and recycling in the city of Tokyo. The overall precipitation is about 1,405mm/year (equivalent to 100%) with an evapotranspiration rate of 412mm/year (29%), while runoff and infiltration account for 634mm/year (45%) and 359mm/year (26%), respectively. Compared with the national average, the runoff water in Tokyo is greater due to the high presence of urbanized areas. The imperviousness is estimated to be greater than 80% within 23 wards. In spite of poor infiltration and limited water availability within Tokyo, human consumption is more than the 1,100mm equivalent rainfall, including recycle water use (199mm) and reclaimed wastewater use (5mm). In Tokyo, there is an artificial water cycle underground composed of the water supply system and the sewerage system, whose water flow is greater than the runoff flow.

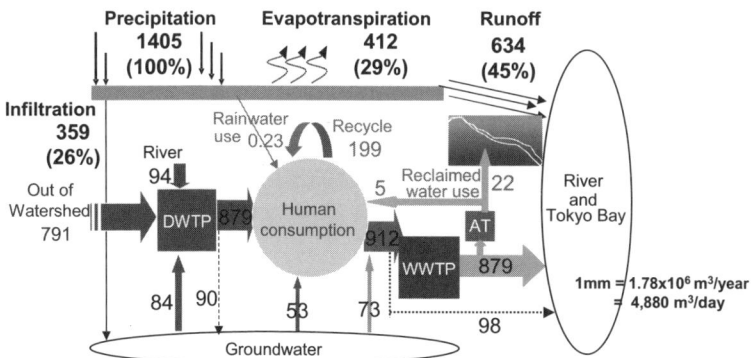

Fig. 2. 6. Water balance and recycling in Tokyo (unit: mm/year)
DWTP: Drinking water treatment plant, WWTP: Wastewater treatment plant, AT: Advanced treatment

2.2 Roles and Functions of Sewerage System

2.2.1 History of the sewerage system and its status in Japan

The first modern sewerage system in Japan was the Kanda sewerage, which was built in 1884 in the Tokyo area. In 1922, the first wastewater treatment plant was built in Mikawashima as shown in Fig.2.7. The construction of a sewerage system to cover the entire Tokyo area did not start in earnest until the end of the World War II.

70.5% of the Japanese population used a sewerage system as of fiscal year 2006 though there is a need to increase the percentage of covered area in small towns with populations below 100,000, as shown in Fig.2.8.

Sewer System in Kanda (Tokyo)   Mikawashima Treatment Plant (Tokyo)

Fig.2.7. The first modern sewerage system and aeration tanks in the first wastewater treatment plant in Japan

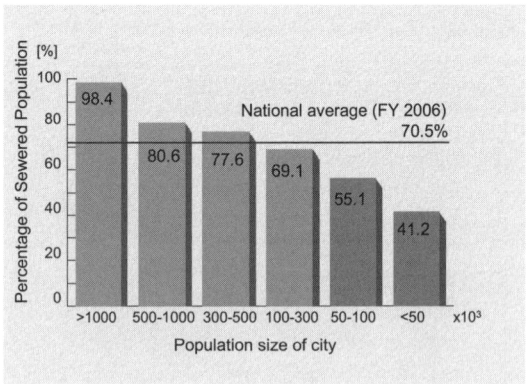

Fig. 2. 8. Percentage of population with sewerage according to population size of cities in Japan

2.2.2 The role of the sewerage system

The role of the sewerage system has been generally divided into the following three responsibilities; i) wastewater collection and treatment, ii) flood and inundation control, and iii) conservation of water quality in public water bodies.

These points have greatly contributed to improved living environments in urban areas. The sewerage system has an important role as an essential element of the urban infrastructure supporting safety, cultural life and urban activities. The construction of sewerage systems has contributed to handling wastewater from households and factories and providing a flush toilet system. In addition, communities are protected from water damage by quickly removing storm water through a drainage system, while the water environment can be improved by the construction and operation of wastewater treatment plants, allowing a decrease of the pollution levels in urban rivers.

2.2.3 Components of the sewerage system

Sewerage systems are made up of the following three components:
- Sewers and drainage

Sewers collect wastewater and transport it to treatment plants. The total length of pipes in the 23 wards of Tokyo comes to 15,000km. Pipes range in size from 200mm to 8m in diameter. The drainage system discharges storm water for flood control. A conceptual illustration of a sewerage system, including a pumping station, is shown in Fig.2.9.

- Pumping stations

Water collected in the sewers passes through pumping stations where it is sent to wastewater treatment plants. Sewer pipes are set at a slope to allow wastewater to naturally flow downwards, so the sewer can end up rather deep underground. Another important function of pumping stations is to guard against flooding. Pumping stations act to quickly discharge water from heavy rains into bays and rivers.

- Wastewater treatment plant (WWTP)

Sewage is generally treated by an activated sludge process in WWTPs. The configuration of the process is illustrated in Fig.2.10. For example, the Tokyo Metropolitan Sewerage Bureau manages 20 plants that treat around 5.8 million cubic meters of wastewater everyday. Since the activated sludge treatment of wastewater concomitantly produces waste sludge, treatment of the excess sludge is also required.

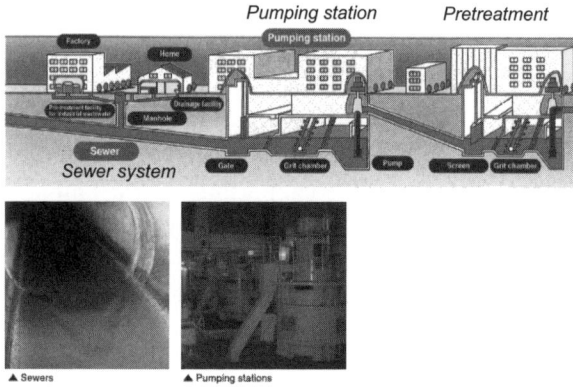

Fig. 2. 9. Sewerage system and pumping station

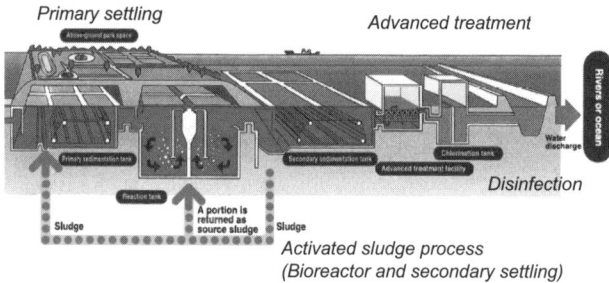

Fig. 2. 10. Activated sludge treatment process with advanced treatment

2.2.4 Wastewater treatment systems

The conventional activated sludge process has been applied for sewage treatment in many large cities. The sequential batch reactor (SBR) system is popular for small size treatment systems, and is included in the category of others in Fig.2.11. Nowadays, the oxidation ditch (OD) process has become widely applied for small sized plants. Different treatment configurations of OD and SBR are shown in Fig.2.12.

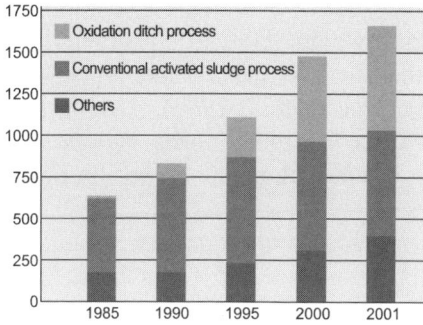

Fig. 2. 11. Number of wastewater treatment plants for different types of treatment systems.

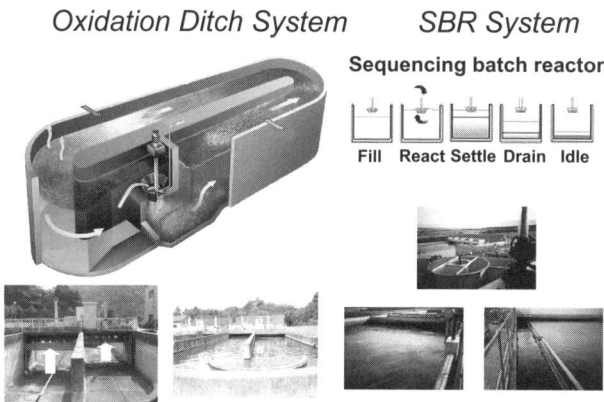

Fig. 2. 12. Oxidation ditch system and Sequential batch reactor system.

2.3 New Roles and Mission of Urban Sewerage System

2.3.1 Rainwater infiltration

The sewerage system in Tokyo has been basically designed to cope with a heavy rainfall of 50mm/hr. The large impervious surface areas and the limited flow capacity of urban rivers have made the storm water flow management system extremely important. Being a water conservation conscious city, the Tokyo Metropolitan Sewerage Bureau constructed a new type of sewerage system in 1980s that incorporates various types of infiltration facilities over a 1400-hectare area (Fujita, 1984). The infiltration facilities include infiltration soakaways, trenches, curbs and permeable pavements. Fig.2.13 illustrates the basic scheme of an infiltration-type sewer system.

The construction of infiltration facilities mainly aimed at inundation control in rapidly urbanized areas by reduced storm water peak flows. It lessened the necessity of improvement works of urban rivers. These infiltration facilities provide secondary benefits such as the recharge of groundwater and the reduction of non-point pollutant loads from urban surfaces. The groundwater recharge is considered essential to secure a sound water cycle in Tokyo. Although Tokyo has restricted and controlled groundwater use, recharged groundwater is a potential storage of water resources and water can be withdrawn in the future if necessary.

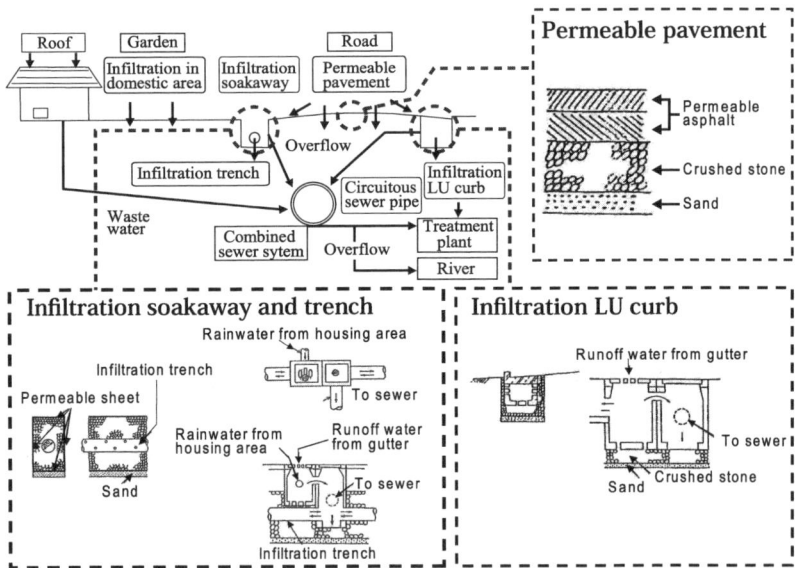

Fig. 2. 13. Configuration of Infiltration-type Sewerage System

2.3.2 Rainwater harvesting

A good example of rainwater harvesting can be seen in the National Sport Stadium in Tokyo, as shown in Fig.2.14. It was introduced in 1984 and was a frontier attempt at large-scale harvesting. The system utilizes a tank with an effective capacity of $750m^3$ to store the rainwater collected from the roof (with an area of $8400m^2$). The collected water is used as a cooling system and toilet flushing. Other examples are the City Office Building in Sumida ward, which introduced a rainwater harvesting system in 1988, Tokyo Dome (1988) and Fukuoka Baseball Stadium (1999).

Fig.2.15 shows the rainwater use at the Tokyo Dome with a capacity of $3000m^3$. The maximum capacity for storage is controlled at $2000m^3$. This means that one-third of its capacity is used for flood control (left empty for this case), another $1000 m^3$ is used for miscellaneous water use and the remaining $1000 m^3$ is for emergency water use such as during earthquakes and fires. In this building, rainwater is used in combination with reclaimed wastewater.

Fig. 2. 14. Rainwater harvesting at the National Sport Stadium

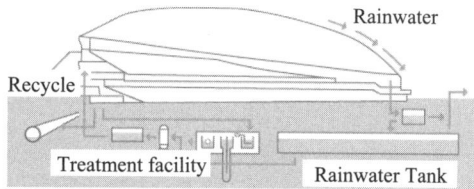

Fig. 2. 15. Rainwater tank for flood control and harvesting at Tokyo Dome

2.3.3 Use of Reclaimed Wastewater

The water amount of about 16 billion m^3 is annually supplied for domestic use in Japan. Domestic wastewater from 70% of the total population in Japan flows into public sewerage systems. In the 2003 fiscal year, sewage of 13.7 billion m^3 was treated and eventually discharged into rivers, lakes, and bays. Not all treated wastewater is discharged into bodies of water. As much as 186 million m^3 of treated wastewater is reused. This amount corresponds to 1.4% of the total amount of treated wastewater. Although the percentage is not very high, sewerage systems play an important role in the water cycle and contribute to effective water use in urban areas. In other words, reclaimed wastewater is a precious water resource and its efficient use should be promoted in case of limited availability of water.

Fig.2.16 shows the different uses of reclaimed water. The environmental use, landscape use and snow melting use are three major uses at the nationwide level. Reclaimed water has been also utilized for recreation, agriculture, industry, flushing toilets and so on.

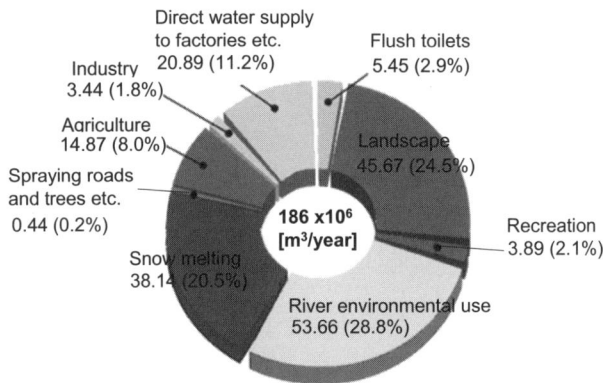

Fig. 2. 16. Reclaimed wastewater use in Japan (Fiscal year of 2003)

A famous example of toilet flushing use can be found in the skyscrapers in the Shinjuku area. This area has a higher demand of water than usual, as shown in Fig.2.17. The treatment plant supplies reclaimed wastewater with disinfection to storage tank in a water recycle centre. This means that they have a dual system where it is possible to use the supply water and the recycled water.

New types of reclaimed wastewater uses have been introduced as well, such as in flush toilets. Examples are splashing water on water-retaining asphalt roads for mitigating the heat island phenomenon, non-point pollution control and train washing water, as shown in Fig.2.18.

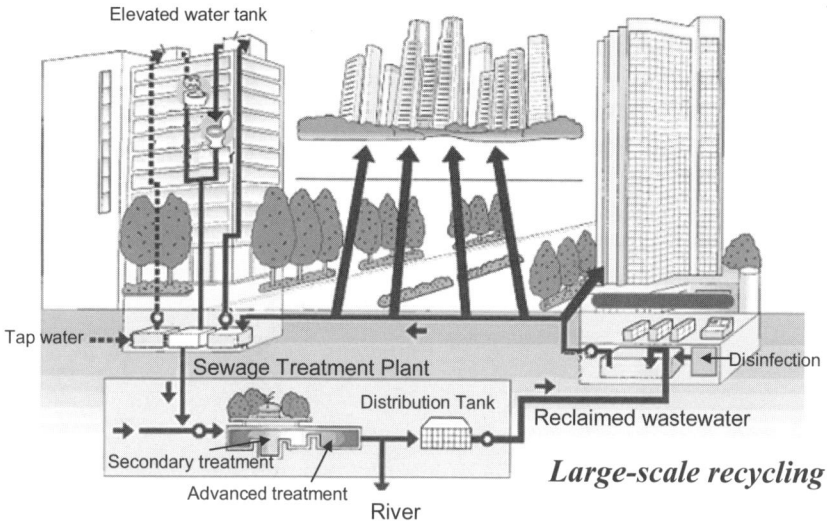

Fig. 2. 17. Reclaimed wastewater use in skyscrapers in Shinjuku, Tokyo

Road splashing water for heat island control and non-point pollution control

Car washing water

Office building district using reclaimed water for toilet flushing

Fig. 2. 18. Reclaimed water use for miscellaneous purposes

The concept of wastewater reclamation can also be used to restore the flow of urban rivers and water channels. The restoration project supplies reclaimed water to urban rivers whose base flow has decreased due to high urbanization and high numbers of impervious surfaces. As part of the solution, reclaimed water is used to replenish urban river flow. Tama River was the main source of drinking water during the Edo era since 1654 when a water channel was constructed. The river water was introduced to a drinking water treatment plant in the central area of Tokyo until 1965. At that time, most of the intake water was transported to a new drinking water treatment plant in the suburbs of Tokyo, leaving the water channels without sufficient flow. Nowadays, the reclaimed water from the sewage treatment plants is used to revitalize the dried-up water channels.

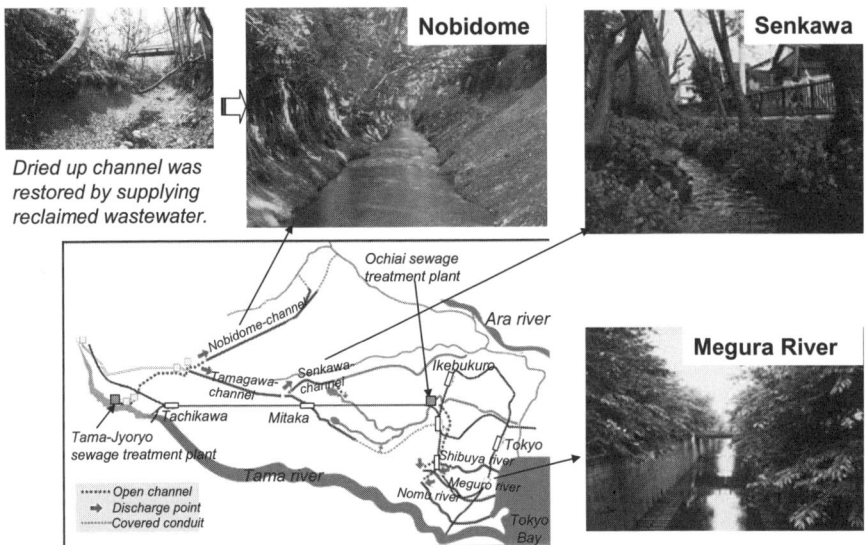

Fig. 2. 19. Reclaimed water use for restoration of water environment and aquatic amenity

It is very important to consider reclaimed wastewater usage not only from the viewpoint of quantity stability, but also quality risk that could rise according to exposure levels to human beings in water use. Therefore, water quality should be evaluated to characterize its potential for reuse and recycling. The Ministry of Land, Infrastructure and Transportation has established and revised the water quality criteria and guidelines for reclaimed wastewater use for miscellaneous purposes since 1981. In April 2005, the water quality standard and facility standard were established for four respective reuses of treated wastewater (Tajima, 2005).

The standards for treated water reuse are presented in Table 2.1. Depending on the final intended use, such as toilet flushing, sprinkling, landscape and recreation, the treatment and quality level can change. For example, chemical precipitation is required for recreational use as well as sand filtration, because higher exposure to human beings is expected for this purpose.

Table 2.1. Standards and Guidelines for reclaimed wastewater quality for miscellaneous use

Usage Parameters	Flushing water	Sprinkling water	Water for Landscape use	Water for Recreational use
Escherichia coli (CFU/100ml)	Not detectable	Not detectable	<1000 (Tentative values)	Not detectable
pH	5.8-8.6	5.8-8.6	5.8-8.6	5.8-8.6
Appearance	Shall not be distasteful	Shall not be distasteful	Shall not be distasteful	Shall not be distasteful
Turbidity (degree)	<2 (Control target)	<2 (Control target)	<2 (Control target)	<2
Chromaticity (degree)	-	-	<40	<10
Odour	Shall not be distasteful	Shall not be distasteful	Shall not be distasteful	Shall not be distasteful
Residual chlorine (mg/l)	>0.1(free) >0.4(combined)	>0.1(free) >0.4(combined)	Not be stipulated	>0.1(free) >0.4(combined)
Facility Standards	Sand filtration or its equivalent	Sand filtration or its equivalent	Sand filtration or its equivalent	Precipitation + Sand filtration or its equivalent

2.3.4 Use of Sludge

Sewage sludge has been regarded as precious biomass, a resource for different material production and has been used in soil improvement, fertilizer, bricks, cement ingredients and road material. In the case of Japan, sludge is mainly used for building material (43%), land reclamation (39%) and agriculture land application (15%). Among them, 12% and 20% of sludge is recycled for fertilizers and cement ingredients, respectively as shown in Fig.2.20.

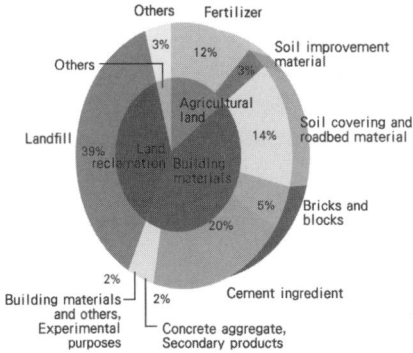

Fig. 2. 20. Effective use of sewage sludge and product forms after treatment

2.3.5 Other possible functions of sewerage systems

Sewerage facility spaces can be used for sports and green park areas, concert halls and network space for fibre optic cable, as shown in Fig.2.21. In addition, treatment plants can be a good learning ground for citizens' environmental education, while open spaces at treatment sites can provide emergency evacuation space after earthquakes.

Concert hall above a pumping station

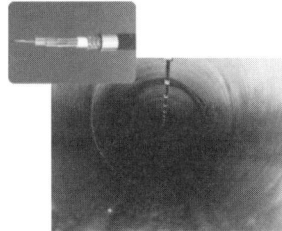

Network cable installation insides sewer pipes

Green park and tennis courts in WWTP site

Gymnastic hall above treatment plants

Fig. 2. 21. Space use in sewerage facilities and inside of pipes

Another example of wastewater use is as a temperature regulator. Heat pumps can be used to utilize heating and cooling functions of wastewater. During the summer, the wastewater temperature is lower than the air temperature, and can play a cooling role while wastewater temperature is higher that the air temperature in the winter, and can be used for heating, as shown in Fig.2.22.

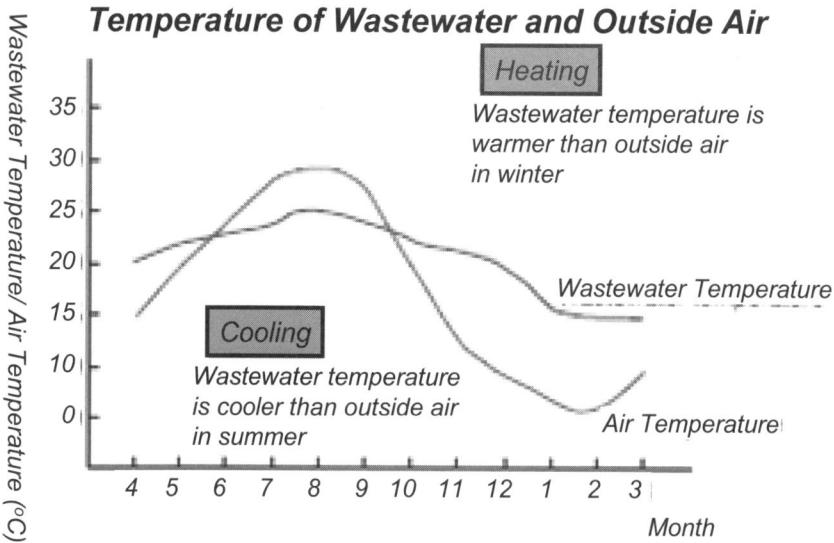

Fig. 2.22. Heat pump facility and seasonal changes of wastewater and air temperatures

2.4 Summary

Tokyo has taken several measures to meet the increased water demand from the viewpoint of sustainable water use. Rainwater and reclaimed wastewater have been applied for various purposes including the restoration of small rivers and waterways. From the viewpoint of human health risk, water quality guidelines are required for the proper usage of reclaimed wastewater.

The application of infiltration facilities for flood control is an example of using groundwater recharge for securing a sound water cycle. It would be possible to provide more potential water resource storage for the future. In addition, infiltration facilities can provide a secondary benefit of the reduction of non-point pollutant loads from urban surfaces.

Sewerage systems can take on diverse roles with the aim of creating a pleasant water environment and recycling society. It is the mission of sewage works not only to protect citizens from flood disaster and inundation damage and to prevent water pollution, but also to promote multiple functions towards Sustainable Urban Development.

References

[1] Fujita S. (1984) Experimental sewer system for reduction of urban storm runoff. Proc. of 3rd Int. Conf. on Urban Storm Drainage, Gutenberg, Sweden.

[2] Furumai H. (2007) Reclaimed stormwater and wastewater and factors affecting their reuse, Chapter 14 in Cities of the Future (Eds.: V. Novotny and P. Brown), 218-235, IWA publishing,

[3] Japan Ministry of the Environment (2002) Water Environmental Management in Japan, http://www.env.go.jp/en/water/wq/pamph/page35-36.html

[4] Japan Sewage Works Association (2002) Making Great Breakthroughs – All about the Sewage Works in Japan, http://www.jswa.jp/en/jswa-en/ http://www.sewerhistory.org/articles/whregion/japan_waj01/index.htm

[5] Population Division of the Department of Economic and Social Affairs of the United Nations Secretariat (2003) World Population Prospects: The 2000 Revision and World Urbanization Prospects: The 2001 Revision, http://esa.un.org/unpp

[6] Tajima A. (2005). The establishment of the new technical standard for the treated wastewater reuse, Proc. of the International Workshop on Rainwater and Reclaimed Water for Urban Sustainable Water Use, Tokyo, Japan. http://env.t.u-tokyo.ac.jp/furumailab/crest/workshop05/june10pm_2.pdf

[7] Tokyo Metropolitan Government (1999) Mater plan of water cycle (in Japanese) Tokyo Metropolitan Government (2005) Sewerage in Tokyo 2005 http://www.gesui.metro.tokyo.jp/english/english.htm

[8] Wagner W., Gawel J., Furumai H., De Souza M.P., Teixeira D., Rios L., Ohgaki S., Zehnder A.J.B. and Hemond H.F. (2002). Sustainable Watershed management: An International Multi-watershed Case Study. Ambio, 31(1), 2-13.

3. Urban River Management: Harmonizing River Ecosystem Conservation

Takeyoshi Chibana

Department of Civil Engineering, the University of Tokyo
7-3-1, Hongo, Bunkyo-ku, Tokyo, 113-8656, Japan

3.1 Introduction

In 1997, river environmental conservation was declared one of the important purposes of river management in River Law. Before this year, the main purposes of river management concerned flood control and water use only. During the 1970s, however, ecological conditions in rivers became worse and worse. At this time, government and citizens started to try to restore and preserve river environments. Against such a background, this chapter focuses on how to make urban river ecosystems harmonize with the mission of flood control.

In Japan, rivers are often quite steep, leading to a sudden rise in the discharge rate after rainfall. Especially in regards to large rivers, Japan has suffered from severe flood damage and the government and citizens have made great effort to control floods. Recently the number of damaging floods has been gradually decreasing, but there are still several problems in river management. Therefore, rivers have been continuously regulated and changed, especially in urbanized areas. For example, the photograph shown in Fig. 3.1 was taken in Gifu prefecture in Japan. All over Japan, we could see this kind of landscape in the past. This river runs through the village and village people have their houses near the river with no barrier between their village and the river. There is easy access to the riverfront.

Fig. 3.1 River that runs through a rural village

Fig. 3.2 Typical urban river

There are many aquatic species found in this river ecosystem (insects, frogs, fish, etc.). In such natural rivers, the interaction between forest and river is also important. Leaves fall into the river and decompose due to the activities of aquatic creatures, such as the feeding activity of macro-invertebrates.

Water quality is affected by decomposed organic matter, and at the same time, river insects move to terrestrial areas during their adult stage and die or are eaten. Material moves from land to river and river to land. Such material circulation takes place in and around rivers. This is the original condition of Japanese rivers.

However, Fig. 3.2 shows one of the typical urban river channels in which both banks are covered by concrete while the channel has many steps. Steps are constructed to reduce the riverbed slope and prevent erosion when a meandering river course has been shortcut and straightened.

These concrete structures are often found in urban rivers. In this case, there is no interaction between the forest and the river and also fish neither migrate upstream nor find their preferred habitat. Around such a river, residents seem not to care about the river, because no fishing activities are found and the water quality is usually bad. In this case, it is difficult for residents to keep their motivation to improve the river.

3.2 The effect of revetment works on the impression of a river

Against such a background, our laboratory conducted a questionnaire survey. Citizens walking along the river were asked about their impression of the river as well as questions related to the bank condition, water quality, vegetation, riverbed condition, and surrounding landscape. First, the favorability rating for each river was quantified, and then the correlation between each factor and the total impression was analyzed. In Fig. 3.3, pictures 1 and 2 are of one river and picture 3 and 4 are of another river. The water quality of 1 and 2 is better than 3 and 4. In 2 and 4, steps to the riverfront are installed on the bank, so people can approach the riverfront. All sites are inside or adjacent to the Asa River basin that will be introduced later.

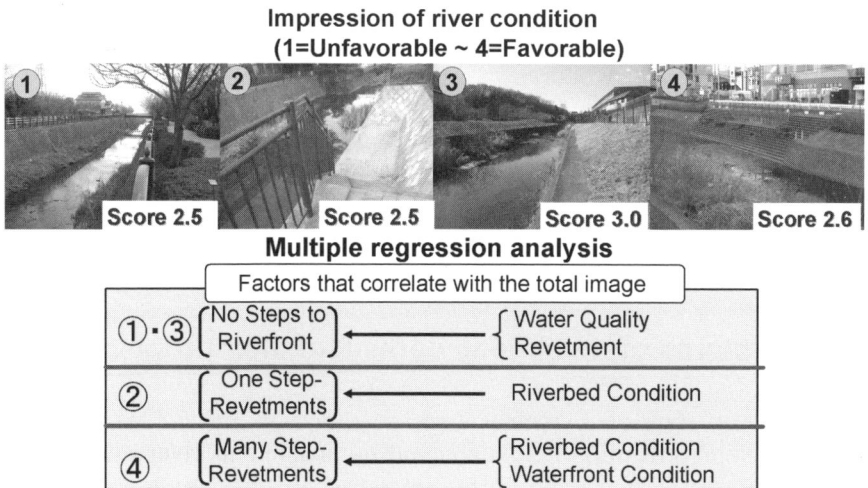

Fig. 3.3 Results of questionnaire survey in urban rivers

As is shown in Fig. 3.3, the favorability rating was 2.5 in three of four sites on a scale of 1 to 4 where 1 is unfavorable and 4 favorable. So a score of 2.5 means just 50% favorability. As this result shows, the condition of the river is not a matter of any consequence for residents around this river. However, in river 3 the score is higher than the other cases, though this river has no steps to the riverfront on the riverbank and the water quality is not so good. Next we analyzed which factors correlate with the total impression. The result is also shown in Fig. 3.3. In the case of photo 1 and 3 where no steps are installed, water quality and revetment (bank protection) were found to highly correlate with the total impression. This result shows that their impressions tend to be good in rivers where the revetment condition and the water quality are good. In the case of photo 2, there is one step and people can approach the riverfront, and the riverbed condition highly correlates with the total impression. For this site, those who answered that the riverbed was dirty or that there was garbage also answered that the river was not good, while people who answered that riverbed condition was good answered that the total image of the river was also favorable. In the case of photo 3, the riverbed condition and the riverfront condition (vegetation, shape of shore etc) correlated with the total impression. This result shows that conditions inside a river channel affect the evaluation of a river's condition.

If there are no steps to the riverfront, citizens evaluated the river conditions based on water quality and revetment because they have never thought about nor approached the riverfront. In that case, the revetment condition and the water quality image were most important. Actually, water quality was not judged based on the actual condition. This analysis clarified that revetment works are very important for the evaluation of citizens when they cannot approach the riverfront. But in order to make citizens take an interest in riverbed condition, step-revetment is necessary in urban rivers. The design of revetment is quite important in urban rivers, so the revetment works in Japan and their effects and problems will be introduced below.

3.3 Structures of revetment works and their problems

The basic structure of revetment works is shown in Fig. 3.4. Usually we consider slope protection that is covered by concrete as revetment. However, the crown of the levee and foot protection are also important parts for bank protection.

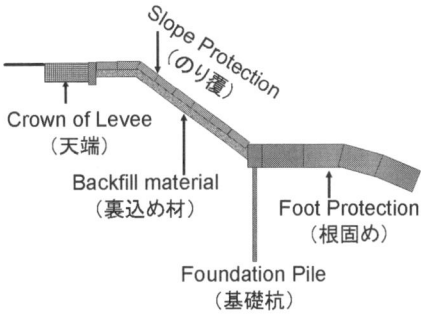

Fig. 3.4 Structure of revetment works
(In the right picture, the foot protection is usually under water level, but it can be observed here due to riverbed degradation.)

Fig. 3.5 Actual revetment damaged by flood before the construction of foot protection and the crown of the levee

Fig. 3.6 Projections to increase the roughness of the surface

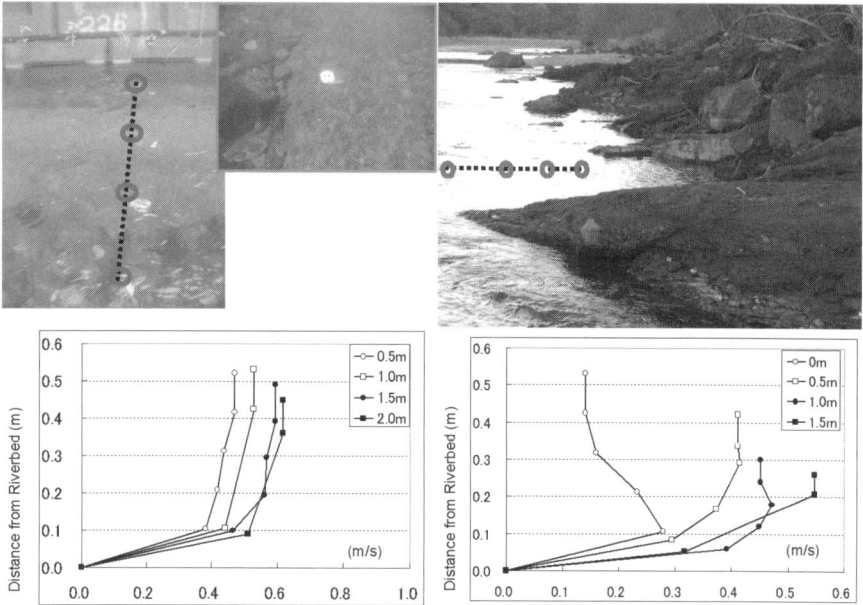

Fig. 3.7 Comparison of the velocity profiles near the river shore
between an artificial (left) and a natural (right) bank

The picture in Fig. 3.5 shows actual revetment works damaged by a flood and demonstrates the importance of the crown of the levee and foot protection. The backfill material behind the concrete that can be observed in the picture is sand and gravel. They were washed out from the top during the flood, because the water level increased before the completion of the crown. In the photo, foot protection is shown under construction along the river channel from left to right. However, before the completion, a very severe flood came. The area with foot protection was safe, but in the right hand side, the foot of the slope protection was damaged. As a result, the surface protection slid out.

It is clear that coverage of the crown and the foot is necessary to protect the riverbank, but these protections as well as the slope protection also damage the river ecosystem and reduce the diversity of river morphology.

One of the problems is monotonous slope protection. As Fig. 3.6 shows, usually some obstacles increase the roughness of the surface in order to reduce the velocity during flood, but the velocity reduction is not very large during normal flow. Fig. 3.7 compares the diversity of velocity profiles near shore between a natural and an artificial bank. Our laboratory conducted field measurements of two rivers. In the figure below, the y-axis shows the distance from the riverbed to the surface and the x-axis shows the velocity. In front of the revetment (left picture), the velocity profiles follow a log distribution. However, at distances of 50 cm, 1 m, 1.5 m and 2 m, the profile is almost constant since depth is completely the same and there is no roughness on the foot protection. In this case, there is neither divergence nor convergence of flow in the cross-sectional direction. But in the natural river without any revetment works (right picture), several kinds of velocity profiles were observed. In some cases, due to the turbulent flow, very high velocity is found near the bottom and very slow velocity near the surface. Thus various velocity profiles are found and create diverse riverbed conditions.

Another problem is caused by foot protections that affect a wider area than the previous case. Fig. 3.6 shows the revetment work in one river. In this site, a topographic survey was conducted in the longitudinal direction along the indicated line. Its result is also shown in Fig. 3.8. If foot protections were installed, the flow power would not attenuate due to the erosion of the riverbank or riverbed during a flood. Then all riverbed materials would be washed out from the bed and the same depth would continue downstream. Usually many small concrete blocks are combined together and are installed as foot protections, because the foot protections should adjust to morphological change (see the picture in Fig. 3.4). However, as Fig. 3.8 shows, the longitudinal variation of depth disappears. In this case, fish cannot choose their suitable depth in a river. Moreover, even if the depth is suitable for a specific fish, the fluctuation of water level degrades the suitability of the pool. It is important to know how to create diverse flow and riverbed conditions from the viewpoint of ecosystem conservation.

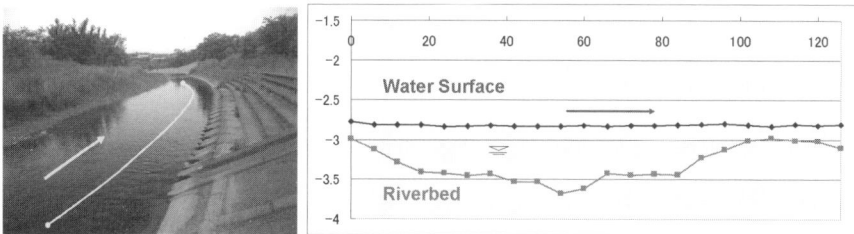

Fig. 3.8 Monotonous riverbed condition along revetment

3.4 Nature-friendly riverbank improvements

Against such a background as mentioned above, companies and governments have begun making improved revetments.

As of 1990, the Japanese government started to restore the river environmental condition. But at the beginning of the restoration program, almost all river works focused only on bank protection. As Fig. 3.9 shows, more than half (60%) of nature-friendly river works conducted at the beginning were improvements of low water revetments. In most cases, citizens criticized concrete blocks for destroying the natural landscape. Also, low water revetments can be easily changed, because they were originally artificial.

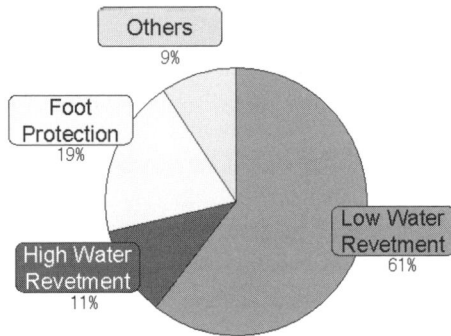

Fig. 3.9 Breakdown of nature friendly river works conducted in 1997

Fig. 3.10 Revetment works using natural stones

Fig. 3.11 Combination of fish nest blocks and vegetation blocks

First, the government started to use stone and rocks for slope protection as shown in Fig. 3.10. This technique of slope protection is called "stone pitching" or "stone masonry." The difference between "stone pitching" and "stone masonry" is according to their slope. If the slope of the bank is steeper than 1:1 (45 degrees) then it is called "masonry." Mildly sloped banks are generally preferred by citizens, but it is difficult to install them in urban rivers. As Fig. 3.3 shows, the preferred site has mildly sloping banks around the top of the revetment. There are two kinds of construction techniques to utilize stones for revetments. One is the "dry stone fill type" and the other is the "filled joint type." In the former case, concrete is not utilized to pile stones, so vegetation can grow among stones and fish can utilize the void for their habitats. Meanwhile, the size of stones should be large enough not to be washed out by floods. So sometimes the landscape does not look like the original condition despite the extra effort. In the later case, small stones can be utilized, but there are no spaces for habitat among stones. Additionally, it is not so easy to install concrete among stones.

Even if concrete material is utilized for revetments, ecosystem conservation can be taken into account in river works. One example is the "fish nest block." The structure is very simple. There are holes and irregularities on the slope protection and fish can hide among these blocks and sometimes enter this block. During times of severe flood, fish can utilize this facility. However, the problem of the fish nest block is that there is no covering above the water level that is usually offered by overhanging vegetation in natural rivers. In such conditions, fish do not want to go into the blocks. Therefore, sometimes vegetation blocks are also installed on the fish nest block as is shown in Fig. 3.11. The combination of these blocks is popular recently. There are voids on the blocks and vegetation can grow. The picture shows fish nest blocks installed at a very shallow and fast area and thus, it remained unutilized.

3.5 Problems of present nature-friendly river works and future visions

Even if several kinds of improved blocks are installed, artificial revetment does not have the same functions as natural banks.Briefly speaking, in installing revetments, the following aspects are missing:

- A characteristic environment that depends on the geological conditions
- Longitudinal transition of riverbank topography
- Dynamism of river topography

| Gravel (Cut bank) | Gravel (Mild bank) | Rock & Boulder |

Fig. 3.12 Natural riverbanks that depend on their geological conditions

Fig. 3.13 Monotonous condition in flow direction

First, the natural condition is quite different among rivers, and each river has its own function. For example, the shape of a riverbank depends on the geological conditions, as is shown in Fig. 3.12. In a mountain region, the riverbed is composed of rocks or boulders, so we can find a quite complicated shoreline, as is shown in the right picture of Fig. 3.7. Diverse flow conditions are found around rocks and boulders. The shape of these rocks is not easily changed by flood. In this case, the area around the rocks was scoured and a very big pool was created. We observed many fish and insects in the area. In an alluvial plane, the riverbank is composed of sand and gravel and is easily eroded. Therefore the slope of riverbank is different between the convex and the concave bank. Steep banks provide a high environmental gradient. For example, fish can stay near shore where the velocity is low and easily access the swift flow area in order to catch drifting material. On the other hand, the mild bank provides a shallow and slow flow area for juvenile fish, etc.

Second, considerations for longitudinal diversity are usually missing. In Fig. 3.13, the concrete blocks with vegetation were installed, but the riverbank should be almost vertical due to urbanization. In the longitudinal direction, there is no change and the monotonous bank condition continues.

Fig. 3.14 Nature friendly revetments sometimes lose their functions
due to the dynamism of the river
(Upper left picture was offered by Mr. Y. Kimitsuka.)

Moreover, as is shown in Fig. 3.8, the height of riverbed tends to be constant in the flow direction. Usually the planning of the river channel is examined based on the cross-sectional image.

Finally, the dynamism of the river should be considered in the design. Floods and droughts are the common characteristics of natural rivers. The rate of discharge and the water level often change, topography is altered and material on the riverbed is renewed during the high-flow condition. In some cases, floods also destroy the vegetation. However, the revetments shown above are quite stable and there is no consideration for the river's dynamism. As Fig. 3.14 shows, those structures cannot function after several floods. In the case of the upper pictures, an artificially created embankment could be a good habitat or refuge for fish as there is no velocity in this area. However, the same conditions are created during the high flow condition and this area becomes a dead water zone. Then, fine material can collect in this embankment. In the case of the lower pictures, after the construction of a fish nest block, the riverbed elevation was degraded and the block became exposed at water level.

If we observe the natural condition of riverbanks carefully, we can get some hint about proper riverbank protection. We do not only use natural stones, but also have to design both lateral and longitudinal cross-sections of revetments that mimic the regional topography. At the same time, morphological transition should be predicted.

3.6 Flow regime in an urbanized river basin

Fig. 3.15 shows the Shakujii River that runs through Tokyo. In urban rivers, the revetment is like a cliff and the water surface is a long way down from the ground level. The photograph was taken four days after a very big typhoon, so the water level was much higher than the usual condition, but it is still quite lower than ground level. Even if steps are installed, as shown in the picture, it is not very easy to get to the riverfront. The important point is why this kind of structure is necessary in urban rivers. In order to examine the reason, we have to focus on the characteristics of the discharge in urban rivers. Therefore we have to examine this topic from the perspective of river basin management.

Fig. 3.16 shows the flow regime of the drainage basin of two adjacent rivers that are the branches of the Tama River that runs through Tokyo. One is called the Aki River and the other is called the Asa River. Their drainage basin areas are almost the same. Within the Asa River basin is Hachioji city and rapid urbanization occurred in this region around 1970. Now, 43% of the whole river basin is covered by artificial structures such as houses, buildings and so on.

Fig. 3.15 Large capacity of an urban river channel

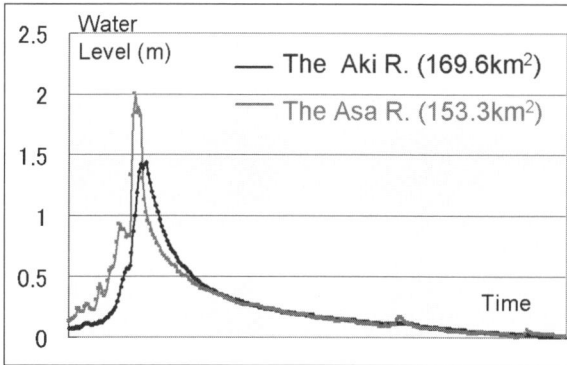

Fig. 3.16 The flow regime of urbanized and relatively natural branches of the Tama River

Fig. 3.17 Quick response of discharge in the urbanized river

If we compare the flow regime of these two catchments, we can easily understand the difference of characteristics. After rainfall, the water level in the Asa River suddenly increases and becomes very high, but then suddenly decreases. In the case of Aki River, which is in a relatively natural condition, the rise and decrease of water level is gradual. This flow regime means that the retention capacity of these two basins is completely different. Hachioji city is covered by concrete and thus the rainfall cannot infiltrate the ground. But in the case of the Aki River, the drainage basin extends towards the mountain region and is covered with trees and grassy areas.

In order to understand this condition, I will introduce two pictures in Fig. 3.17 taken in the Asa River during fieldwork for university students. The site is near the river mouth of the Asa River. We began the fieldwork on a fine day. However, around 3:30pm, it began to rain suddenly. Intensive rainfall lasted for about 40 minutes. At this time, field measurements were cancelled. After that, around 4:30pm, on returning to the area, we found the water level had increased about 1 m. This is a typical characteristic for an urbanized river. In the case of the Shakujii River, the same situation often happens and the water level rises and falls suddenly. Thus, a quite deep channel was created. At the same time, this kind of flow characteristics means that if precipitation is limited, the river channel easily dries up due to no infiltration into the ground. Fig. 3.18 is also taken in a branch of the Asa River.

3.7 Basin scale countermeasures

Discharge management is quite important in the urbanized region from the viewpoint of ecosystem conservation as well as flood control. Of course, we have to reduce the peak discharge, but at the same time we have to maintain a certain water level even in a dry season. In Japan, many dams have been constructed to maintain water level. However, in the case of very small streams in the urbanized region, a big dam cannot be constructed and moreover, dams are not preferable structures in terms of ecological conservation.

Fig. 3.19 shows the picture of the present Asa River basin and the traditional landscape in hilly areas. Originally, there were several valleys in a hilly area that contained wetlands and small streams. Traditionally the wetland is utilized as paddy fields. But recently, such regions have been changed into residential areas that cannot retain water. As the right picture in Fig. 3.19 shows, the traditional paddy fields can retain water during rainfall and also support a diverse ecosystem. Ideally, the environment shown in the right picture of Fig. 3.19 should be conserved or restored.

Fig. 3.18 A branch of an urbanized river without water

Fig. 3.19 Urbanized hilly area in the Asa River basin (left)
and traditional paddy fields in a hilly area in Japan (right)

Fig. 3.20 Countermeasures to retain water in a drainage basin:
permeable pavement in a parking area (left) and a retarding basin (right)

Even in the urbanized area, it is important to retain water in a drainage basin. Fig. 3.20 shows some measures. The left picture shows permeable pavement in a parking lot and the right one shows a retarding basin in a hilly area. In addition to these measures, leeching pits are often installed in a yard of a house to make water infiltrate into ground.

As mentioned, impermeable pavement has altered flow regimes in rivers, therefore countermeasures such as those in Fig. 3.20 are necessary. However, another important factor that has altered flow regime is the urban river channel itself. If we compare Fig. 3.1 and Fig. 3.2 again, we can understand the difference of riverbed roughness. In natural rivers like the one shown in Fig. 3.1, the riverbed is composed of gravel and grasses. Additionally, the river's course is meandering and its slope is gentle. Under such conditions, the water level tends to increase during floods and sometimes the water overflows, and therefore the river channel and surrounding area retain water during a flood.

Fig. 3.21 Two-layered river

On the other hand, the river channel in Fig. 3.2 is designed to discharge water quickly and reduce the water level as much as possible. Then the downstream area collects a large amount of water at the same time from many branches. The regulation of discharge becomes more difficult in large, urbanized rivers. From the viewpoint of drainage basin management, some small branches should have more retention capacity inside their channels and contribute to delaying the timing of peak discharge if the surrounding area is not densely populated.

3.8 Importance of floods in a river

In the previous section, it was mentioned that the retention of water in a drainage basin or branches contributes to both the reduction of peak discharge and the maintenance of low flow discharge. But if no floods occur in a river, the capacity of a river channel can be reduced and people can access the riverfront much easier. In that case, a flood way seems to be effective for discharge regulation. Based on this idea, a two-layered river is created in places. In a two-layered river, the floodway is constructed under the original river and then people can walk along the riverside. During heavy rainfall, excess water runs through the underground flood way. Fig. 3.21 shows examples of actual two-layered rivers and their structure. The distance from the walking area to the water surface is very close. Under the river a big floodway is installed. So even during heavy rainfall, water can pass through the underground channel. It seems that nature friendly river works can harmonize flood control.

Fig. 3.22 Renewal of riverbed material by floods

Floods can renew the riverbed condition, as shown in Fig. 3.22. When no floods come, fine materials collect on the riverbed and algae overgrows. This kind of situation is often found in stable discharge conditions, such as the upper layer of the two-layered river. If floods come, these fine materials are washed out and clean gravel can be found in the river.

Another problem is the succession of river ecosystem. If there are no disturbances, those who can go ahead of the competition can thrive. Then the diversity in a river decreases. This situation can be explained by a very simple equation. This equation is called a Logistic Equation and is shown below:

$$\frac{dN}{dt} = rN(1 - \frac{N}{K}) \tag{1}$$

where "r" is the rate of increase, "N" is the number of individuals, and "K" is the carrying capacity.

This equation is used to predict the number of individuals. For example, if we take an area with no vegetation and put in one species, we can estimate how the area will be covered with vegetation. Based on this equation, the number of trees, for example, increases rapidly at first, but due to the second term (1-N/K), the increasing ratio will gradually decrease. Finally, the number of individuals will correspond to K. This equation is utilized for the case of one species, and does not account for interaction among several species.

However, if we consider competition among several species, the Lotka-Volterra Equation shown below is utilized:

$$\frac{dN_1}{dt} = r_1 N_1 (1 - \frac{N_1 + \alpha_1 N_2}{K_1}) \tag{2}$$

$$\frac{dN_2}{dt} = r_2 N_2 (1 - \frac{N_2 + \alpha_2 N_1}{K_2}) \tag{3}$$

The structure of the equation is almost the same as the Logistic Equation, but the term α is a competition coefficient. The number of N_2 affects the increasing ratio of N_1 and vice versa. Then, the number of trees is simulated by using these equations. In the simulation, it was assumed that there were two species: pioneer (N_1) and competitor (N_2). The pioneer is the species that can grow rapidly even in a barren area. So the increasing ratio r_1 is large. In the case of competitor, the increasing ratio r_2 is not so large but they can go ahead of the competition. In the case of competition between these two species, the number of trees will change as shown in the left of Fig. 3.23. Initially, both two species increase, then the N_1 species will decrease and ultimately will become extinct due to the competition. The N_2 species then attains a stable number. This kind of competition is often found in actual rivers without disturbances. Ultimately, the river channel will be covered by the stronger species. This is called succession. Thus the diversity of the ecosystem is diminished.

However, if we assume disturbance as shown in the right side of Fig. 3.23, the trend becomes different. In that case, the pioneer species will be able to live for a long time and the competitive species will not be able to cover the whole area. As this simulation shows, several disturbances can retain the high diversity in natural conditions. This is also true in actual cases and if there are floods, then many species can live in the river.

Thus, artificial floods are often caused by releasing water from a dam. The purpose is the improvement of the river environment's condition and to stimulate the diversity of aquatic creatures. Another important purpose of the artificial flood is to restore river topography. Artificial floods can remove fine deposited particles and gravel and create a sand beach at the riverside.

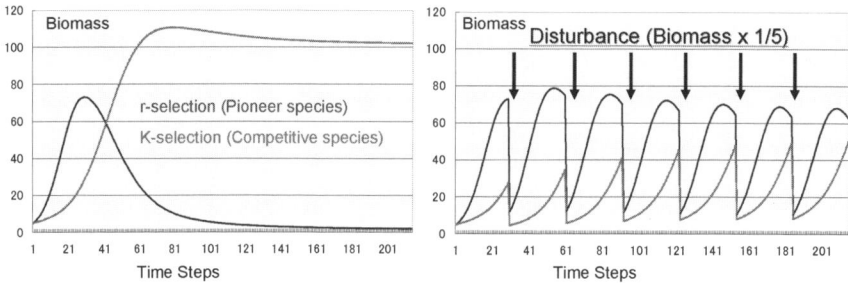

Fig. 3.23 The behavior of the Lotka-Volterra Equation

3.9 Perspectives on urban river management harmonizing with river ecosystem conservation

In this chapter, measures for river ecosystem conservation have been introduced on several scales. Ideally, a river without any artificial structures is the best condition for aquatic creatures, but this is impossible in urbanized areas. In urban rivers, the design of revetments is quite important. In order to make local citizens take an interest in the river, some approaches to the riverfront are essential. Otherwise, people estimate the river condition by focusing just on revetments. Anyway, the improvement of revetments is essential.

In order to mitigate the negative effect of revetments on aquatic creatures, we have to mimic natural bank conditions that are regionally specific. The diversity of the revetment type in the flow direction is very important, but is not often taken into account. Moreover, a prediction of topographical change is necessary when revetments are installed.

For appropriate design of river channels, discharge regulation is an important issue. However, we have to focus on the whole drainage basin for discharge management. Sometimes, two-layered rivers or dams are effective countermeasures, but floods are also important events that sustain the diversity of aquatic creatures. The key point in discharge regulation is how and where storm water should be retained. Conservation of rice paddies and the installation of permeable pavement are important, but at the same time retention capacity inside a river channel should be improved. In some places, the capacity of the river channel can be increased in order to store water during floods and delay the timing of peak discharge.

4. Development of Coastal Urban Regions with Sustainable Topographical Environments

Shinji Sato

Department of Civil Engineering, The University of Tokyo
7-3-1, Hongo, Bunkyo-ku, Tokyo, 113-8656, Japan

4.1 Importance of Water Management for Human Beings

Water is one of the most important elements in the ecological system. Since the pre-historic age, human beings have tended to live near water due to the convenience in water usage for their daily needs. It is true that most of the ancient civilizations began in areas enriched with water resources. This implies that sustainable water management has been one of the essential keys in human society's development. On the other hand, however, the history of this development is also one of struggle against natural disasters due to water. Besides all the privileges that we attain from water, we also have to remember the disastrous aspect of water and manage it well to achieve a safe and comfortable society.

Recently the world has experienced two historic natural disasters. The first was the Indian Ocean Tsunami on December 26, 2004, which devastated coastal settlements around the Indian Ocean and killed more than 200,000 people. The other was storm surge due to Hurricane Katrina on August 31, 2005, which inundated New Orleans and nearby areas in the United States. In Japan, frequent attacks of typhoons generate extraordinary floods, such as the Kurokura River Flood in August 2001 and the Maruyama River Flood in October 2004. Heavy rainfall also triggers debris flows in mountainous region, carrying huge amounts of sediment together with the water. In order to utilize the benefits of water in human society without suffering the damage that water can bring, such disastrous aspects of water behavior should be understood.

Japan is considered to be prone to various natural disasters, owing to its geographical and geological conditions. Fig.4.1 shows the number of deaths due to natural disasters in Japan over the past 60 years (Disaster Prevention White Paper 2007). The horizontal axis represents the time af-

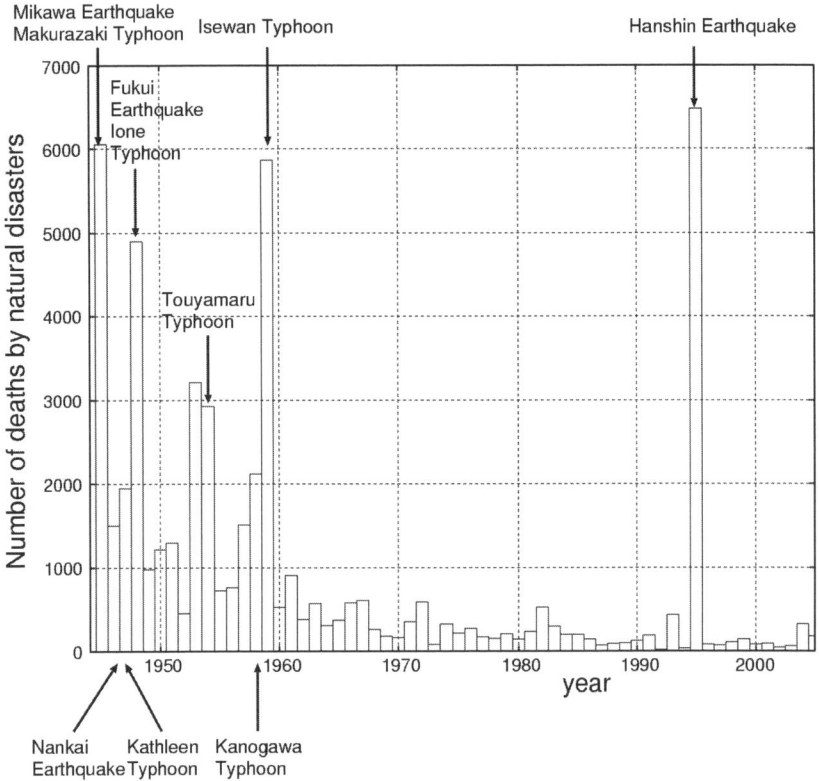

Fig. 4.1. Number of deaths due to natural disasters in Japan

ter the end of the World War II and the vertical axis represents the number of deaths by the natural disasters. A large number of deaths can be seen during the post-war period. This is due to two possible reasons. The first is the frequent occurrence of large earthquakes and typhoons in this period. The second is the lack of disaster mitigation management, as well as of financial support for land management. Meanwhile, it should be noted that the post-war investment in disaster mitigation has possibly increased the safety level and successfully decreased the number of deaths. The details of the number of deaths due to various disasters (Disaster Prevention

White Paper 2007) indicates that earthquakes produce an extraordinary number of victims but their frequency is relatively low. It is also noted that the damage of earthquakes is more locally concentrated than other natural disasters. Meanwhile natural disasters due to wind, flood and snow are more frequent, although the number of deaths in any one event is not as high as those caused by an earthquake. Although the number of deaths due to wind, flood and snow disasters has decreased owing to post-war investment in disaster mitigation, it still remains at a significant level.

4.2 Coastal Zone Management

As described above, waterfront areas have been among the first zones utilized for the development of cities. Among waterfront areas, coastal areas are of particular importance and many of the world's mega-cities have developed in the coastal zone, such as New York, Shanghai, Tokyo, Osaka and Nagoya. The coastal zone is unique and valuable in the sense that it serves as a buffer zone for various physical, chemical and biological processes among the atmospheric, the aquatic and the terrestrial zones. Various materials are transported within these zones, helping to maintain its sustainability. From an ecological perspective, the coastal zone is a unique and rich ecosystem with various fauna and flora. Without the coastal zone, some species would fail to maintain their life cycles. A typical and famous example is the turtle, which cannot lay its eggs in the sand unless sandy beaches of a moderate size are preserved. In respect to human activities, the coastal zone has many different uses, including tourism, industry and recreation. We have to remember that the environment in the coastal zone is also highly variable and fragile. Waves and currents vary greatly between storm and moderate conditions; winds and temperature can change significantly between day and night. In order to maintain the coastal zone as an attractive and safe zone, we need to develop technologies for disaster mitigation, since the coastal zone is highly sensitive to even slight changes in the environment and is prone to natural hazards such as inundation and erosion by storm waves and tsunamis.

Human activities have increased in scope and scale in the last century. Significant anthropogenic impacts on the coastal zone are observed in many areas, such as sea level rise due to global climate change, erosion due to the imbalance in sediment budget and eutrophication due to the accumulation of nutrients. It is essential to understand characteristics of material flow within coastal zones and develop technologies for utilization, disaster mitigation and environmental conservation in order to keep coastal zones attractive, safe and sustainable.

4.2.1 Coastal Zones and Sea Level

The location of the coastal zone is defined as the intersection of land and the sea level. In the geological time scale, the sea level has varied in the range of 10 to 10^2 m. Fig.4.2 illustrates the temporal variation in sea level. The horizontal axis is time in thousands of years, with zero implying the present. 10,000 years before the present, the sea level was 40m lower than the present level. Meanwhile 6,000 years ago, the sea level was 3-5m higher than the present level. Such variation in the sea level has changed the location and the environment of coastal zones substantially. Fig.4.3 illustrates the shape of shoreline geometry in the Kanto district, Japan, showing the shoreline 6,000 years ago located far inland.

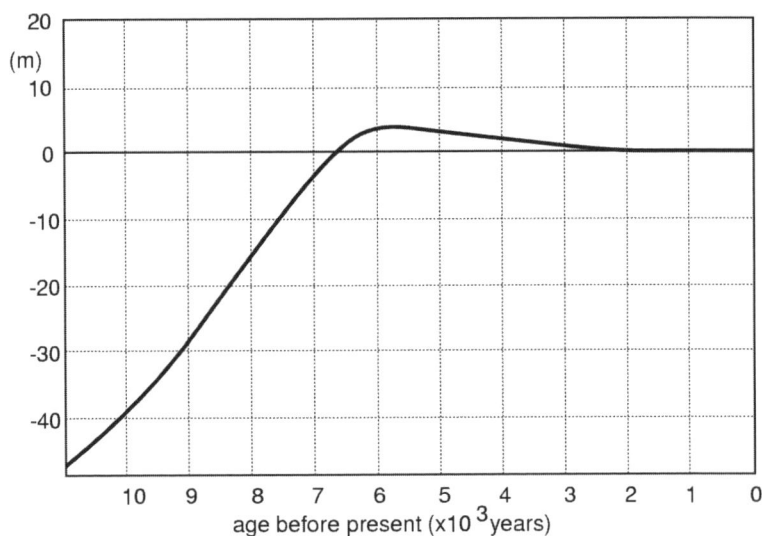

Fig. 4.2. Temporal variation in the sea level (modified from Umitsu 1994)

Fig.4.2 shows that the sea level has been relatively stable during the last 6,000 years. In the modern era, however, it is anticipated that the sea level will rise owing to anthropogenic impacts resulting from excessive emission of greenhouse gases. The IPCC (2007) estimated that the sea level would rise by 18-59cm by the end of the 21st century. If the future prediction trend is drawn in Fig.4.2, the rate of sea level rise is found to be of similar rate as 10,000 years ago. Establishment of a careful monitoring system and a precautious strategy is urgent.

In addition to the sea level rising, drastic changes in the coastal zone environment have been observed due to rapid urbanization. Such changes are classified into two types: those by direct pressure of urbanization near large coastal cities mostly located in large bays and those observed on coasts facing the ocean owing to the rapid deformation of coastal topography. Typical examples will be discussed below.

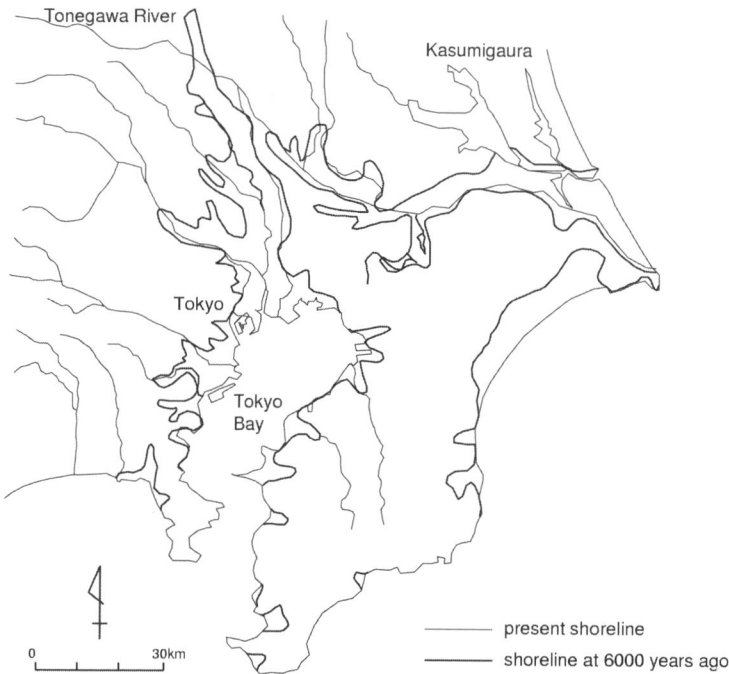

Fig. 4.3. Shoreline geometries in Kanto area, Japan (extracted from Japan Association for Quaternary Research 1987)

4.2.2 Rapid Change in Coastal Environments due to Urbanization

Fig.4.4 shows the topography of the waterfront area of northern Tokyo Bay. The dark area in the upper part represents lowland below sea level. The lowland was created by the alluvial deposits of the old Tonegawa River that flowed into Tokyo Bay until 400 years ago. The Tonegawa River carried significant amounts of sediment, forming the lowland as well as a shallow area in Tokyo Bay. The presence of the shallow area helped the rapid development of the coastal zone around the Tokyo Bay by reclamation. Such reclamation efforts have continued since the Edo era, resulting in a near total reclamation of the area around Tokyo Bay now.

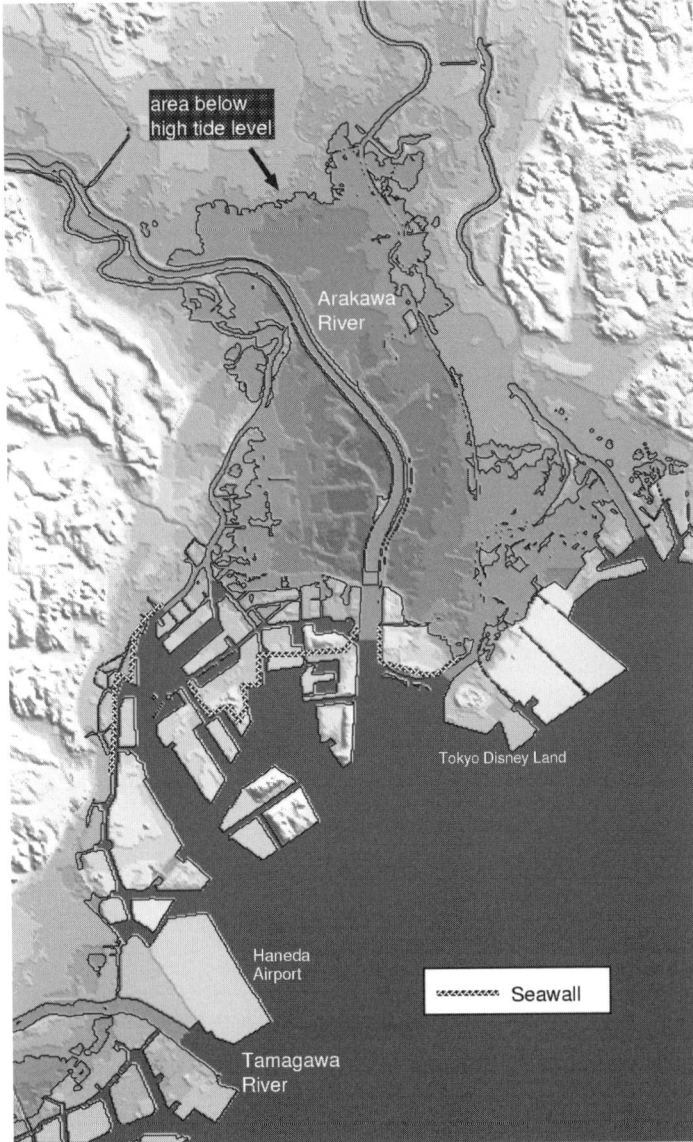

Fig. 4.4. Topography in the coastal zone in northern Tokyo Bay

A typical example of how urbanization changes the bay-front area is found in Fig.4.5, showing the coastal zone near the Tokyo Disneyland located at the mouth of the Edogawa River. The old picture was taken in 1947 and the new picture in 1997. Large areas have been reclaimed over these fifty years, including the land on which Tokyo Disneyland has been built. It should also be noted that the mouth of the Edogawa River is the river mouth of the old Tonegawa River. In the Edo era, some four hundred years ago, the government changed the route of the Tonegawa River in order to avoid flooding in Edo and to enhance water transport. The route was changed to an eastern route flowing into the Pacific Ocean. The route change successfully enhanced development of the waterfront area of Edo, and contributed to the modern urbanization of Tokyo as seen in Fig.4.5. However, the rapid development brought about two risks for the coastal zone; water quality degradation and a greater potential risk for storm surge.

Fig. 4.5. Comparison of the Edogawa River mouth in 1947 and 1997

Water quality degradation

In the shallow zone at the northern part of Tokyo Bay, reclamation was accomplished using sediment dredged from the seabed of the shallow area. As a result, deep trenches were created, in which the seawater stayed with stratification. At the bottoms of the trenches, oxygen is depleted by the nutrient overload from urbanized areas as well as by the accumulation of detritus of phytoplankton. Sometimes the bottom water becomes anoxic with the production of hydrogen sulfide. The anoxic sea water is upwelled by strong northern winds, mostly dominant in early autumn, producing milky blue sea water due to oxidization of hydrogen sulfide and killing fish and benthic animals on the shallow sea bed (Sasaki and Isobe 2000). The presence of anoxic water is observed in a wide area in Tokyo Bay owing to rapid urbanization.

Risk of storm surge

We also have to pay attention to the possibility of disastrous events in the Tokyo Bay zone. For example, when a typhoon generates strong southern winds, the sea level will rise, especially at the northern end of the bay. This is the mechanism of storm surge caused by typhoons or strong winds.

Storm surge is an event where the sea level will be raised by a huge low-pressure system, such as typhoons and hurricanes. There are two mechanisms for the sea level rising in the storm surge. The first mechanism develops because of low pressure in the atmosphere. Atmospheric pressure at the centre of the typhoon is usually lower by several tens hPa compared with the average atmospheric pressure. Thus the sea level will be raised in order to compensate for the imbalance in atmospheric pressure. The decrease in the atmospheric pressure by 1hPa will elevate the sea level by approximately 1 cm.

The second mechanism is related to the wind stress exerted on the sea surface. In order to balance the wind stress, the sea level should be elevated on the downdrift side of the wind. Consider a body of water in a bay exposed to wind as illustrated in Fig.4.6. The sea level rise $\Delta\eta$ at the end of the bay can be calculated by:

$$\Delta\eta = \frac{\tau_s l}{\rho g h}$$

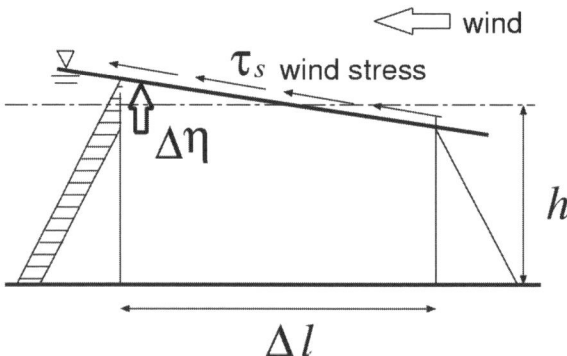

Fig. 4.6. Mechanism for storm surge

where τ_s denotes the wind stress, l is the length of the bay, ρ is the density of seawater, g is the acceleration due to gravity and h is the water depth. As a result, the water level rise $\Delta\eta$ is inversely proportional to the water depth h. Based on this equation, it is confirmed that the presence of shallow areas tends to enhance water level rise significantly. Tokyo Bay has a shallow area at the northern part of the bay, close to the densely populated lowland, where the Tonegawa River carried huge amounts of sediment. The presence of such a shallow area tends to increase the risk of storm surge for the lowland.

4.2.3 Rapid Topographic Change due to Waves and Currents

The other response observed in modern coastal zones is rapid deformation in topography due to waves and currents. Fig.4.7. shows a typical example observed at the river mouth of the Sagami River, Japan, facing the Pacific Ocean. From these pictures we can conclude that the shore is moving toward the land. A phenomenon where the shoreline encroaches upon the land is called coastal erosion, which is a rapid phenomenon compared with the slow retreat due to sea level rise. Since coastal zone management is based on the prediction of coastal zone deformation, the mechanism of deformation should firstly be identified.

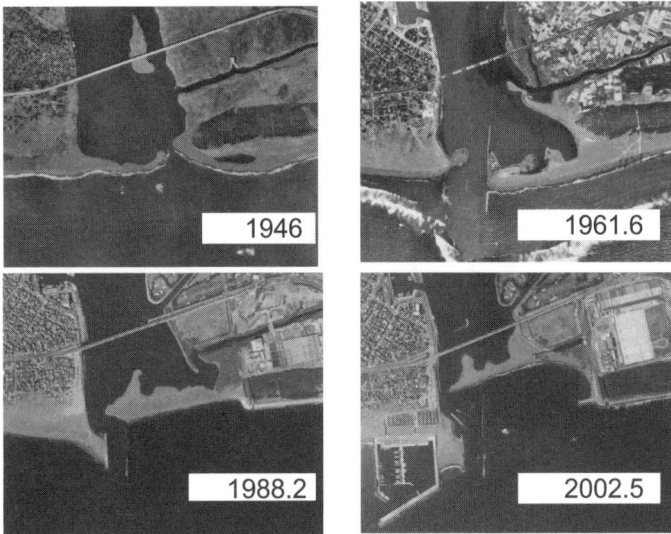

Fig. 4.7. Coastal erosion at the Sagami River mouth

4.3 Prediction of Shoreline Deformation

Coastal zones are subject to chronic deterioration in addition to catastrophic events such as tsunamis and storm surge. One such chronic disaster is coastal erosion. The deformation of coastal topography can be caused by two mechanisms in terms of sediment movement: profile change due to cross-shore sediment movement and bathymetry change due to longshore sediment movement. Profile change is commonly caused by the varying directions of cross-shore movement of sediment by storm waves and mild waves. The shoreline will be eroded by storm waves since they transport more sediment from the shore to the sea. However, the eroded shore will gradually recover as successive mild waves carry sediment back to the shore. Such processes are repeated by the seasonal variation of waves and are mostly found to be reversible.

The change in the nearshore bathymetry in a large region is caused by longshore sediment transport. On the coast where the shoreline is aligned at an oblique angle to the dominant wave direction, longshore sediment transport will be dominant in the direction of the incident wave angle. In contrary to the variable direction of cross-shore transport, the direction of the longshore transport is less variable since the angle of dominant waves commonly varies only within a narrow range. Therefore, coastal deformation due to longshore transport is considered to be an irreversible process in contrast to the reversible profile change due to cross-shore transport. The deformation in the nearshore bathymetry developed by the longshore transport is thus considered to be more important in the sense that it results in permanent deformation of coastal topography.

A shoreline model can predict the shoreline deformation due to the longshore transport. The model is based on the estimation of the longshore transport rate and several assumptions on the nearshore bathymetry change. Consider a beach with parallel depth contours exposed to obliquely incident waves as shown in Fig.4.8, where y is the alongshore coordinate and x_s denotes the shoreline position. Among numerous formulas to estimate the longshore sand transport rate, the widely used CERC formula predicts the longshore sand transport rate by the following equation:

$$I = K P_l \tag{1}$$

where I is the longshore sand transport rate expressed by the immersed weight of sediment. The relationship between the immersed weight transport rate and the volumetric transport rate is expressed by:

$$I = (\rho_s - \rho)\, g\, (1 - \lambda_v)\, Q_y \qquad (2)$$

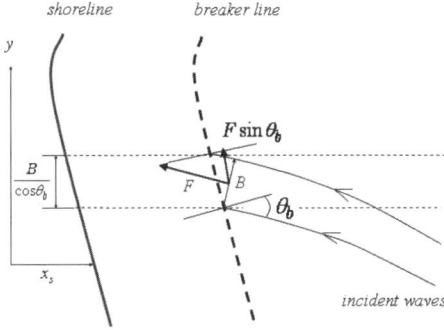

Fig. 4.8. Energy flux of oblique waves incident to coast with parallel depth contours

where ρ_s and ρ are densities of sediment particles and sea water respectively, λ_v the void ratio of the sediment and Q_y is the volumetric longshore sand transport rate including the volume of the void among sediment particles.

On the right hand side of Equation (1), P_l represents the longshore component of the incident wave energy flux per unit length of the shore expressed by:

$$P_l = 1/8 \; \rho g H_b^2 \; c_{gb} \sin \theta_b \cos \theta_b \qquad (3)$$

where H_b denotes the breaker height, c_{gb} the group velocity at the breaking point and θ_b is the wave angle at the breaking point. The coefficient K (=0.77) has been calibrated by laboratory and field data (Komar and Inman 1970). The equation is based on the simple assumption that the work consumed by the transport of sediment in the longshore direction is proportional to the longshore component of energy flux of incident waves. Since the incident wave energy flux F at the breaking point is estimated by:

$$F = 1/8 \; \rho g \, H_b^2 \; c_{gb} \,, \qquad (4)$$

the longshore component of the energy flux will be expressed by $F\sin\theta_b$. Since the waves are obliquely incident at the breaking point, the energy flux within two wave rays with spacing B will be incident to the shore of length $B/\cos\theta_b$. Therefore, the alongshore component P_l of the wave energy flux incident to the unit length of shore will be expressed by Equation (3). On the other hand, the work required to transport a certain amount of sediment in the longshore direction would be proportional to I when the friction coefficient between the bed and the moving sediment particle is assumed to be constant, which leads to the relationship between I and P_l simply expressed by Equation (1).

Then, consider the nearshore bathymetry change due to the longshore sand transport. If the longshore sand transport rate Q_y is constant along the shore, the nearshore bathymetry will be unchanged although sediment is transported. Such a condition is called dynamic equilibrium. However, as illustrated in Fig.4.9, the spatial variation in the longshore sand transport will result in the deformation of the coastal topography. In Fig.4.9, consider a narrow zone with alongshore length Δy. Since the incoming sand volume within time Δt is expressed by $Q_y\Delta t$ and the outgoing sand volume by $\left(Q_y + \partial Q_y / \partial y \bullet \Delta y\right)\Delta t$, the sand volume loss over time Δt will become

$$-\frac{\partial Q_y}{\partial y}\Delta y\Delta t \tag{5}$$

By the conservation of the mass of sediment, this amount should be equal to the volume of the bathymetry change. Assuming that the cross-sectional profile of nearshore bathymetry moves horizontally only in parallel translation, the volume change due to the translation can be related to the change Δx_s in the shoreline position as:

$$h_s\Delta x_s\Delta y \tag{6}$$

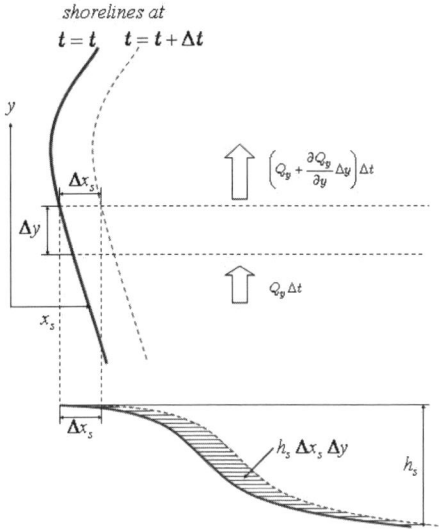

Fig. 4.9. Longshore sand transport and shoreline deformation

where h_s is the height of the littoral zone where sediment transport takes place. By equating Equation (5) and Equation (6) and writing it in a derivative form, the shoreline deformation can be expressed by:

$$h_s \frac{\partial x_s}{\partial t} = -\frac{\partial Q_y}{\partial y}. \tag{7}$$

When the longshore sand transport rate Q_y is expressed by Equation (1) and when the nearshore wave field is uniform, the right side of Equation (7) is expressed in terms of the shoreline change as follows:

$$\frac{\partial Q_y}{\partial y} \propto \frac{\partial}{\partial y}\left(\frac{1}{2}\sin\theta_b\right) \approx \frac{\partial}{\partial y}\tan\theta_b = \frac{\partial}{\partial y}\left(-\frac{\partial x_s}{\partial y}\right) \tag{8}$$

since the wave angle at the breaking point is usually very small. Therefore, the governing equation of the shoreline model to predict the deformation of the shoreline will finally be expressed as:

$$\frac{\partial x_s}{\partial t} = \alpha \frac{\partial^2 x_s}{\partial y^2} \tag{9}$$

where α denotes a parameter representing the wave height, the wave angle and the height of the littoral zone. Equation (9) indicates that the shoreline deformation is simulated by a one-dimensional diffusion equation.

4.4 Coastal Erosion and Its Countermeasures

Fig.4.10 illustrates the shoreline deformation near the river mouth computed by the shoreline model. Consider the deformation of the shoreline of an initially straight beach around a river mouth, where sediment is supplied by a constant rate. When the sediment supply is 300,000 m^3 per year, the shoreline model predicts the formation of delta topography around the river mouth. The shape of the delta is dependent on the rate of sediment supply and the incident wave conditions. For a typical condition of incident wave height of 0.6m, the delta will gradually develop in 1000 years as shown in Fig.4.10. In 1000 years, the topography of the delta almost reaches equilibrium where the longshore sand transport by waves almost balances the sand supply from the river mouth. If the sediment supply from the river is 100,000 m^3 per year, the convex shape of the delta will become smaller as shown in Fig.4.11. This is because the equilibrium shape of the delta is decided by the balance between the sand supply from the river and the longshore transport of sand by waves.

The dynamic response of coastal topography is revealed by considering the shoreline deformation near the river mouth. Fig.4.12 illustrates the shoreline deformation near the river mouth in 40 years starting from the shoreline at 1000 years in Fig.4.10. Since the sediment movement at the time of 1000 years nearly reaches equilibrium, the change in the shape of the shoreline is insignificant. In this situation, most of the sediment supplied from the river is efficiently transported longshore by the convex shape of the delta topography developed over 1000 years. On the other hand, in Fig.4.13, a rapid shoreline retreat is resulted when the sediment supply is decreased suddenly. In this figure, the sediment supply is assumed to decrease to 100,000 m^3 per year suddenly at the time of the 1000th year. Then the shoreline starts to retreat rapidly from the region close to the river mouth as shown in Fig.4.13. The shoreline retreat in 20 years reaches as much as 50m. The speed of erosion can become very rapid once sediment movement becomes unbalanced. Such rapid decrease in sediment supply from the river has been observed in many rivers in Japan, as well as in the world, owing to the rapid growth of human activities in the twentieth century.

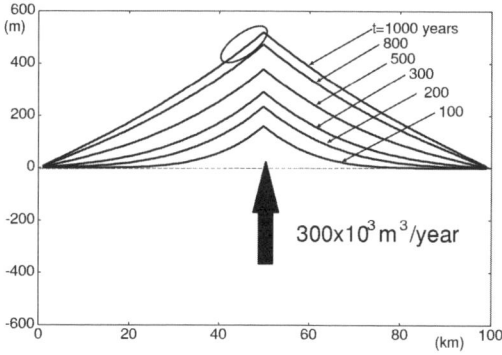

Fig. 4.10. Shoreline deformation around a river mouth (300,000 m³/year)

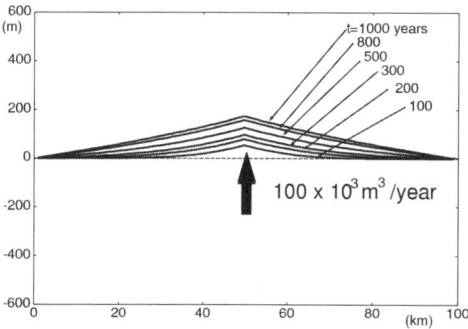

Fig. 4.11. Shoreline deformation around a river mouth (100,000 m³/year)

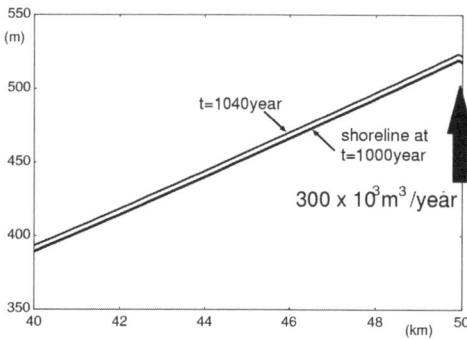

Fig. 4.12. Shoreline deformation around a river mouth for quasi-equilibrium condition (300,000 m³/year)

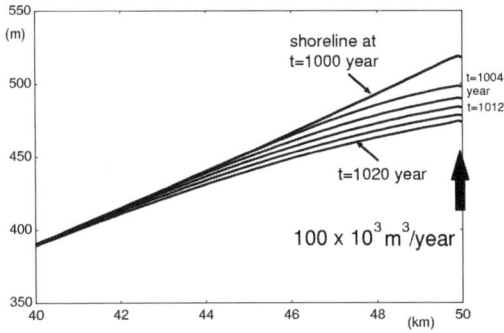

Fig. 4.13. Shoreline deformation around a river mouth for decreased discharge (100,000 m^3/year)

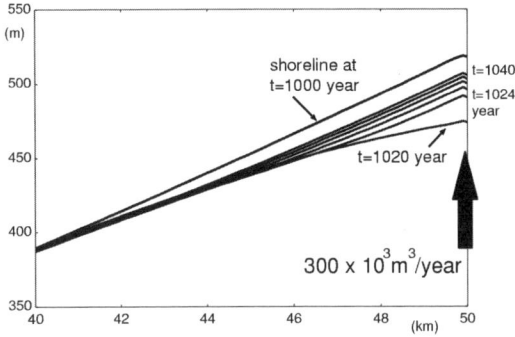

Fig. 4.14. Shoreline recovery after sand supply increase (300,000 m^3/year)

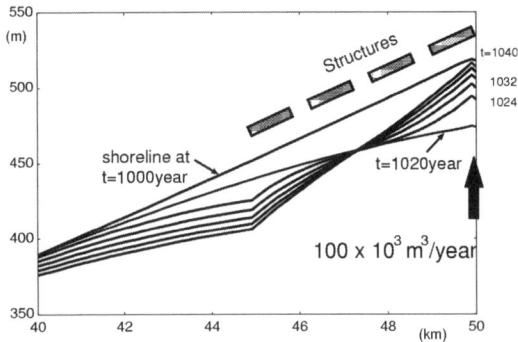

Fig. 4.15. Shoreline deformation due to structures (100,000 m^3/year)

Rapid coastal erosion resulted from the rapid decrease in sediment supply can be mitigated by two countermeasures. The first countermeasure is by an increase in the sediment supply. Fig.4.14 illustrates the shoreline response in 20 years by the sediment supply of 300,000 m^3 per year recovered from 100,000 m^3 per year. Shoreline recovery starts from the river mouth and gradually propagates to the downdrift side. The second countermeasure is by a decrease in sediment movement along the shore. This can be achieved by the construction of shore protection structures that dissipate wave energy in the surf zone where longshore sand transport dominates. Fig.4.15 shows the shoreline response computed for 20 years where shore protection structures are installed at t=1020 year and successfully contribute to decreasing the longshore sand transport rate by one third so as to match the decreased sand supply of 100,000 m^3 per year. Fig.4.15 shows that the shoreline near the river mouth recovers after the introduction of structures. However, the erosion on the downdrift side of the series of structures is accelerated. This is because the effect of structures is limited to the local region where the structures are installed.

Such sudden decrease in the sediment supply from the land to the ocean can be observed in many countries. Sivitsky et al. (2005) reported that the modern sediment delivery from land to ocean has decreased by 1.4 x 10^9 ton per year, which is about 10% of the total amount of natural sediment movement in world rivers. The construction of dams has trapped about 10% of sediment that would otherwise be discharged from land to the ocean. Consequently the erosion of beaches is becoming a significant issue worldwide. The impact of sediment supply decrease on coastal erosion is especially significant where the original sand supply was at a high level. According to Milliman and Meade (1983), sediment delivery to the ocean is large in the new volcanic zone, that is, the Pacific Rim zone and the Alps to the Himalaya zone. In particular, in East Asia and South-east Asia, a huge amount of sediment is discharged into the ocean because of geological conditions and heavy precipitation.

In Japan, the rate of sediment delivered from land to the ocean is as much as 200 million m^3 per year, which is large compared with the small land area of Japan. The large amount of sediment delivery from land to ocean in Japan can be confirmed by the presence of siltation in the reservoirs. Siltation is especially significant along the tectonic line in the central mountainous district. Like most countries worldwide, Japan is facing severe coastal erosion. The rate of erosion was investigated by the comparison of old and new maps, which revealed that the recent speed of erosion was 1.6 km^2 per year in total. The volumetric loss of sediment in the nearshore area, equivalent to the erosion rate of 1.6km^2 per year, amounts to 8 million m^3/year. Although the loss of sediment is considered to be large, the amount is only 4% of the total sediment delivery in Japan. Therefore, mitigation countermeasures based on sediment movement control in the watershed is considered promising.

4.5 Future Visions

4.5.1   Structure or Nourishment?

Fig.4.14 and Fig.4.15 indicate that the construction of shore protection structures or the nourishment, artificial supply of sand to coast, can be considered as effective countermeasures to coastal erosion. However, the selection of countermeasures must be based on the social, natural and geological conditions of each country. In Japan, for example, the rate of sediment production from land to ocean is very large and the population is dense in the area very close to the shoreline. In the Unites States, on the other hand, sediment supply is relatively small especially on the east coast and the population concentration in the coastal zone is less significant. It is quite natural that optimum countermeasures to coastal erosion are different depending on the situation.

Fig.4.16 shows the trend of total shore protection structures in Japan (Coast Statistics, Ministry of Land, Infrastructure and Transport, 1960-2002). Basically, shore protection was initiated by efforts to control waves and currents by structures. In the U.S., however, countermeasures by nourishment have increased and replaced those by structures since the 1960s. This is partly due to the increasing demand for the preservation of natural beaches and due to the natural conditions of relatively small sediment supply in the country. Nourishment can be easily applied since marine sediment is mostly confined inside the nearshore area. Therefore coastal zone management in the sense of sustainable topography in the U.S. can be based on sand management without introducing structures. In the case of Japan, however, where sediment production is larger and the population is denser in the narrow plains close to the coast, huge amounts of sediment are required for the compensation of coastal erosion if only nourishment is considered as the countermeasure. This explains why the optimum shore protection method is different in Japan and in the U.S. However, even in Japan countermeasures to increase the sand supply to the coast should be considered in order to achieve long-term sustainability.

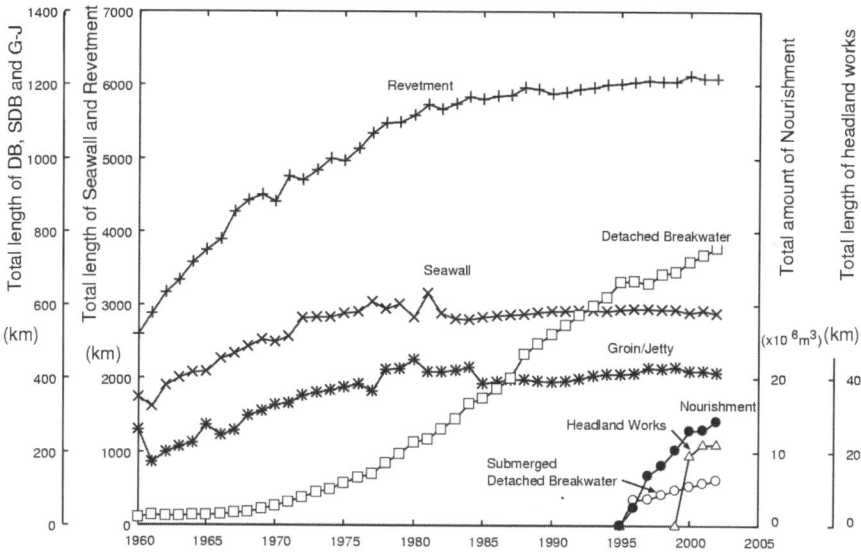

Fig. 4.16. Shore protection structures in Japan

4.5.2   Regional Sediment Management

In order to increase sediment supply from the land to the sea, sediment movement should be managed at the regional scale. Regional sediment management has been standard worldwide ever since the relationship between coastal erosion and various anthropogenic impacts was recognized in the last 50 years. In Japan, as was described in the previous chapter, it is highly expected that mitigation countermeasures in the mountains may consequently lead to an increase in the sediment production from land to the ocean. Such activity has been initiated in the Tenryu River watershed, located in the central mountainous area in Japan. Sediment production in this area has historically been quite large but has decreased significantly due to the construction of a series of large dams, such as the Sakuma Dam in 1951 which led to severe coastal erosion starting from the river mouth.

Fig.4.17 shows the siltation of the reservoir of Sakuma Dam. The dam is located at point 0, and the original riverbed elevation is indicated by a thick solid line. It is noticed that siltation is progressing gradually and now it is almost approaching the dam. The rate of the siltation amounts to 1.3 million m^3/year. Various countermeasures are being discussed in order to recover the continuous sediment flow to the downstream river and to the coast, such as the dredging from the reservoir and the construction of a sediment bypassing tunnel or a sediment discharge gate on the dam. Various countermeasures are also being considered at the watershed scale. Construction of open-type sabo dams is one such effective mitigation countermeasure. However, all efforts should be introduced in a gradual manner but continuously, since the sediment movement at a regional scale is a long-term phenomenon with much inertia, as was described in the previous chapter.

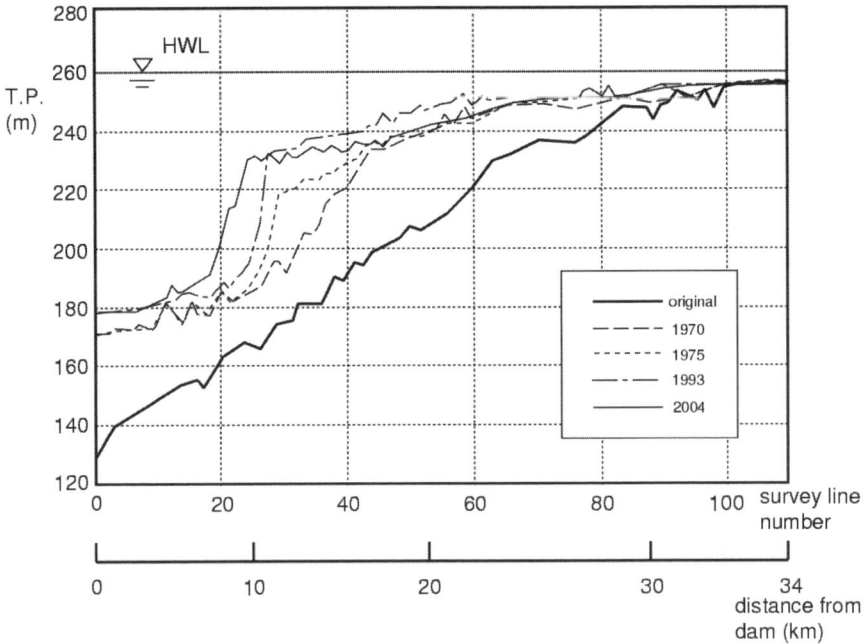

Fig. 4.17. Siltation in the Sakuma Dam reservoir

References

[1] IPCC (2007) WGI: Summary for Policymakers, Climate Change: The Physical Science Basis, http://www.ipcc.ch/

[2] Japanese Government: Disaster Prevention White Paper (2007) http://www.bousai.go.jp/hakusho/h19/index.htm

[3] Japan Association for Quaternary Research(1987) Quaternary Maps of Japan,

[4] Komar, P. and Inman, D.L.: Longshore sand transport on beaches, J. Geophys. Res., Vol. 75, pp. 5914-5927.

[5] Milliman, J.D. and Meade, R.H. (1983) World-wide delivery of river sediment to the oceans, J. Geol., 91, 1-21

[6] Ministry of Land, Infrastructure and Transport: Coast Statistics, Japanese Government, 1960-2002.

[7] Sasaki, J. and M. Isobe (2000) Development of a long-term predictive model for baroclinic circulation and its application to Blue Tide phenomenon in Tokyo Bay, Proc. 27th Int. Conf. on Coastal Eng., Vol. 4, ASCE, pp. 3870-3883

[8] Syvitski, J.P.M., Vörösmarty, C.J., Kettner, A.J. and Green, P.(2005) Impact of humans on the flux of terrestrial sediments to the global coastal ocean, Science, Vol. 308, pp. 376-380

[9] Umitsu, M.(1994) Late Quaternary Environment and Landform Evolution of Riverine Coastal Lowlands, Kokin Shoin, 270p

Part II   Material Flow and Risk

5. Municipal Solid Waste Management for a Sustainable Society

Kazuo Yamamoto

Environmental Science Center, The University of Tokyo
7-3-1, Hongo, Bunkyo-ku, Tokyo 113-8656, Japan

5.1 Sound Material-Cycle Society and 3R

We are in the era of global environment. It is recognized that global warming may lead to a crisis of ecological systems and a shortage of natural resources, resulting in fatal damage against human activities. The concept of a "Low Carbon Society" has been proposed to adapt to global climate change and to make society sustainable by significantly reducing the emission of carbon dioxide as a result of our fossil fuel dependent activities.

A "Sound Material-Cycle Society" is also our concern, where waste minimization is comprehensively achieved by promoting the "3R" (Reduce, Reuse, and Recycle) process, as illustrated in Fig.5.1. Reduce means any effort at production and consumption points to reduce the amount of waste generated. It is noted that the waste and by-products of one manufacturer can be resources to another manufacturer. Reuse points to the use of things repeatedly, including a long use of repaired goods or their parts. Recycle means the conversion of waste to resources, including material recycling and recycling by energy recovery. It is emphasized that the primary focus should be on reducing, then reusing and then recycling. Furthermore, proper disposal should not be neglected.

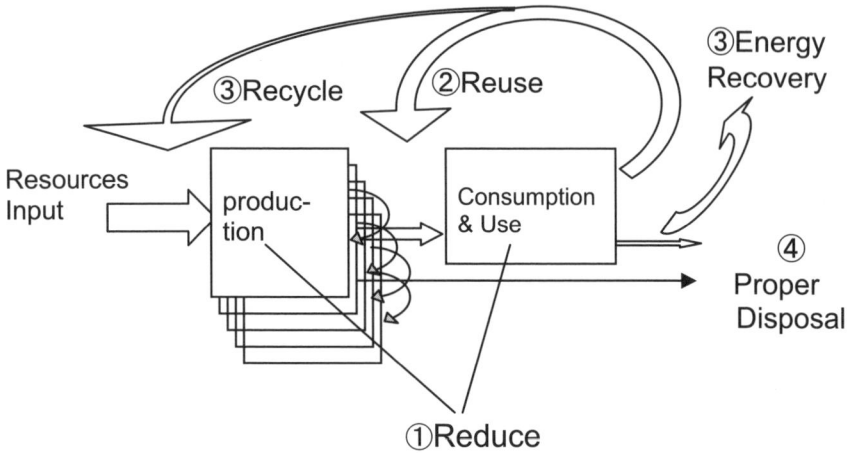

Fig. 5.1. Promoting the 3Rs for a Sound Material-Cycle Society

Whatever the level of waste minimization is, waste will still be produced. Therefore, the proper disposal of solid waste remains essential in promoting the "Sound Material-Cycle Society."

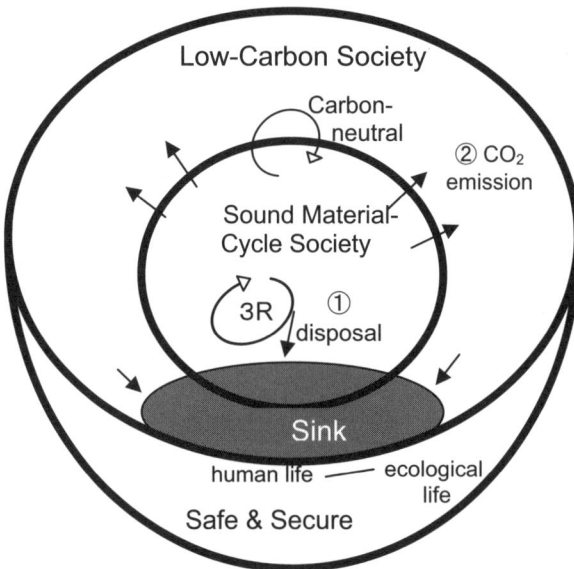

Fig.5.2. The conceptual relationship between the Low Carbon Society and the Sound Material-Cycle Society

The 3R activities must contribute to establish the "Low Carbon Society" as well. Fig.5.2 depicts the conceptual relationship between the "Low Carbon Society" and the "Sound Material-Cycle Society". While it is necessary to establish the "Low Carbon Society" globally, the "Sound Material-Cycle Society" can be locally considered. Both the global and local society must be founded on the basis of safe and secure life of the society member

Waste minimization is usually targeted at the minimization of 'disposal' (Fig.5.2). If the boundary of a sound material-cycle society is the nation, for example Japan, the annual amount of 'CO_2 emission' in Fig.5.2 is much larger in weight than that of 'disposal': the former was about 1,200 million tons/y while the latter was about 35 million tons/y as of FY2004 (Ministry of Environment, 2007). Too much effort to achieve so called 'Zero Emissions' of solid waste may result in further burden to the globe through an increase in the emission of gaseous wastes such as carbon dioxide. Therefore, it is important to minimize waste considering the total emission outside.

Developing countries drive rapid growth of population and some of them are experiencing rapid economic growth also. These enlarge waste problems due to mass production, consumption and use of natural resources, and consequently lead to the environmental deterioration that developed countries have already experienced. Although differences must be recognized in the situation of waste problems between developing and developed countries, it is still worthwhile to consider some common issues from the lessons of the past and present problems encountered in developed countries.

5.2 Municipal Solid Waste Management - Case of Japan

5.2.1 A brief summary of the modernization of municipal solid waste management

During the Edo period (1603-1868) in Japan, a nearly complete recycling system existed, as it was a typical feature of pre-industrial societies. For example, human excreta were valued as fertilizer; kitchen garbage was utilized as animal feed; straw was used after the rice harvest as fuel and raw materials for articles for daily use in farming households. Although the Edo city has densely populated, the surrounding rural area was capable of recycling the waste produced in the Edo city and the cleansing of the city streets was well controlled by the Edo community.

After the Meiji Revolution (1868), industrialization eventually caused problems of waste disposal in urban areas like Tokyo. In 1900, the first waste management law (the Dirt Cleansing Law) was established to ensure sanitation in cities to prevent the spread of infectious diseases. In this law, it was clearly stated that incineration should be considered as much as possible. Furthermore, the ash produced by the incineration was usually sold to farmers as fertilizer (Mizoiri 2007).This economic exchange became another incentive for the use of incineration as a modern sanitation technique. This initiative formed the foundations for the establishment of a management system for municipal solid waste.

After World War II, rapid economic growth brought about a tremendous increase of municipal solid waste until the first oil shock crisis (1973). In 1954, the Public Cleansing Law was enacted, aiming at improvement of public health and the environment by establishing a municipal solid waste management system that focused on sanitary disposal of garbage and human waste. However, the issue of solid waste generation was beyond the Law's expectations and the various industrial activities had caused serious environmental pollution. This situation led to the creation of relevant laws and policies for environmental pollution control and waste management, introducing the "Polluter Pays Principle" (PPP) in which, in cases of pollution, the polluter should pay the cost of the appropriate countermeasure or remediation.

As the waste generated from industrial sectors was much larger than that of municipal waste, it was recognized that the proper management of the industrial waste was of primary importance. Therefore, the Waste Management Law replaced the Public Cleansing Law in 1970, clarifying the responsibilities of businesses on their industrial waste treatment, based on the PPP. In order to specify the responsibility of the waste generator, waste was classified into industrial and general waste.

Table 5.1 shows the up-to-date definition and classification of waste.

Fig. 5.3 shows the changes in the amount of municipal solid waste generated in Japan. During the bubble economy (1985-1990), its generation significantly increased, then stabilized at around 50 million tons a year. Industrial waste generation stabilized at about 400 million tons a year in the 1990s. Recently, the 3R activities for the creation of a material-cycle oriented society made a turning point to reduce the municipal solid waste generation since 2000.

The modernization of municipal solid waste management progressed during the rapid economic growth period. For waste collection, for example, packer-trucks replaced manual collection. 24-hour continuous-loaded large incinerators were introduced in big cities with advanced off-gas treatment as well as energy recovery, while a large number of 8-hour batch-loaded small incinerators were installed in many small municipalities as well. As for improvement of landfill technology, semi-aerobic sanitary landfills were equipped with impermeable liner sheets and the collected leachate treatment process became a standard landfill method for dealing with municipal solid waste.

Table 5.1. Definition and classification of waste (adopted and modified from a report by the Clean Japan Center, 2007)

Definition: things no longer used by their owners nor sold to others, such as refuse, bulky refuse, burnt residue, sludge, excreta, and other wasted solid or liquids or unnecessary things. Radioactive waste and waste polluted by radioactivity are excluded.
Industrial Waste
Waste generated from all kinds of business activities
coal cinders, sludge, waste oil, waste acid/alkali, waste plastics, waste rubber, metal scraps, glass/concrete/ceramic scraps, slag, rubble, soot dust
Waste generated from specific business activities
-waste paper (constructors, paper manufacturers, and book binders)
-wood chips (constructors and lumber mills)
-waste textile (constructors and textile mills)
-animal and plant residue (food processing)
-animal solid waste (slaughter houses)
-animal manure (livestock farms)
-dead animals (livestock farms)
Processed material for the disposal of the above mentioned industrial waste
Imported waste excluding industrial waste listed above, flight and voyage waste, and hand-carried waste
Specially Controlled Industrial Waste (Specified hazardous waste/flammable oil/strong acid and alkali/infectious medical waste and others)
General Waste: waste other than industrial waste
Municipal solid waste
Human excreta (night soil and sludge generated from on-site domestic wastewater treatment process (Jokaso))
Specially Controlled General Waste (specified due to its nature, including toxic, explosive, infectious, etc.)

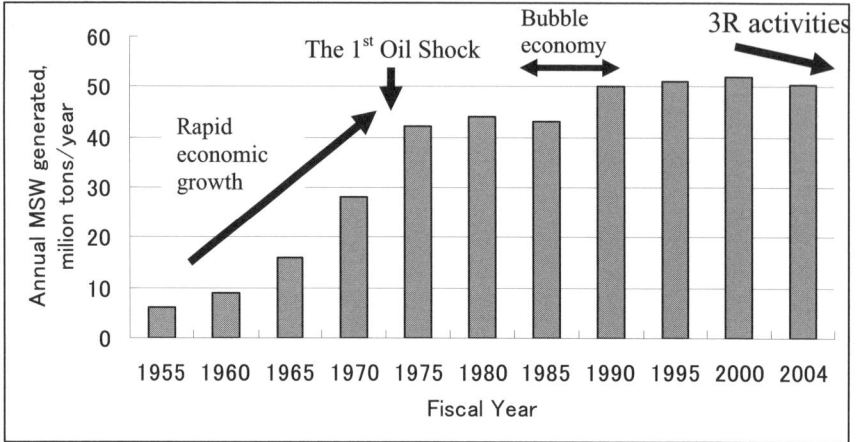

Fig. 5.3. Changes in the amount of municipal solid waste (MSW) generated in Japan.

Note: The figures for 1970 and before are calculated from the amount of the daily waste generation in the special cleansing districts based on the Public Cleansing Law. The data are adopted from a governmental report (Ministry of Environment, 2006).

Japan is among the countries with the highest incineration rates of municipal solid waste, with levels of around 77% as of FY2004. Reasons include a limited habitable land area (about 30% of the territory), and the high temperatures and humidity are non-preferable conditions for organic waste handling. Therefore, there has been a need in Japan for waste volume reduction treatment that is both rapid and sanitary.

5.2.2 Public awareness and acceptance

An example for the social problems associated with waste management is Tokyo's "Waste War" episode. In 1971, Koto Ward was the recipient of the new Yumenoshima landfill site. In previous years, neighbours of various landfill areas reported problems with increases in bird and insect populations, mostly due to the presence of the uncovered waste piles during daily landfill operation. These antecedents created opposition to the presence of a new landfill site and people became more sceptical about where this waste was originated. Therefore, Koto Ward refused to accept the solid waste from Suginami Ward, where residents opposed the construction of a new incineration plant (Suginami Incineration Plant) to treat the waste originated in their own ward. It took nearly three years to compromise each other. As an outcome of this event, the problem of waste treatment and disposal, associated with the "NIMBY" syndrome (Not In My Back Yard), became more obvious in society. People began to think about the enormous volume of waste generated and the necessity to introduce reduction and recycling of waste.

Another example of the social problems created by waste disposal is the vicious circle of improper disposal, "NIMBY-ism," and shortage of landfill sites. The lifestyle based on mass production and mass consumption created an unexpected increase in waste volume during the high economic growth period and the consequent bubble economy period. This garbage boom, together with land limitations and its rocketing price, led to a crisis of the final disposal capacity in the country and led to so-called "illegal dumping." Illegal dumping, the improper management of landfill sites as well, caused environmental deterioration, including groundwater contamination and soil pollution. Public awareness of this issue made people nervous regarding the final disposal. Even though a well-controlled landfill facility was planned, many local residents opposed it as a result of the "NIMBY" idea, which led to difficulty in finding a new landfill site. This again induced improper disposal, leading to a vicious circle.

In order to eliminate illegal dumping, regulatory measurers were strengthened several times by amending the Waste Management Law, mainly in the 1990s. The "Manifest" system was fully introduced to industrial waste, in which a document called 'Manifest' is used to identify all processing steps of the waste treatment until its final disposal, and the responsibility of waste-generating businesses to control their waste output was strengthened throughout the treatment process, including collection, transportation, intermediate treatment and final disposal. The "Outline of the Action Plan towards the Eradication of Illegal Dumping" was publicized in 2004. The action plan includes improvement of public awareness against littered wastes in the neighbourhood, increasing transparency of the waste treatment system, human resource development supporting the system and so on. The establishment of the Illegal Dumping Hotline and a tracking GPS system are named as examples. The "Electronic Manifest system" for industrial waste has been enforced as well.

It is emphasized that the reduction of the final disposal is especially important, in addition to the efforts to stop the vicious circle at all stages, such as enforcing stringent regulations against improper disposal, encouraging good practices in final disposal businesses, gaining public acceptance by employing secured landfill technology, and so on. In fact, if we compare the situation of the shortage of industrial waste landfill sites in FY1998 to that in FY2003, the nationwide average remaining lifetime for landfill sites rose from 3.3 years (FY1998) to 6.1 years (FY2003). The 1998 figure indicated a rather serious shortage of landfill sites for industrial waste, compared to the landfill sites for municipal solid waste, which had 12.2 years of remaining lifetime (FY1998). The difference was mainly due to a significant reduction of the final disposal amount of industrial waste, from 58 million tons (FY1998) to 30 million tons (FY2003). Although industrial waste generation remained about the same at about 400 million tons, the 3R activities in the industrial sector contributed to the reduction of the final disposal.

The reduction of waste through intermediate treatment is also essential. Incineration technology has a great contribution in this respect. The construction of incineration plants for municipal solid waste was practiced from the late 1950s until the 1980s. In the early 1980s, the dioxin problem raised public concern. After that, it became more difficult to get public acceptance for the construction of a new incineration plant. National and local governments took effective action to cope with the dioxin problem by setting new regulations and guidelines for countermeasures against dioxin. Consequently, dioxin emissions all over Japan were drastically reduced, as shown in Fig. 5.4.

Nowadays, dioxin emissions from municipal solid waste incinerators are controlled to a minimal level. There is also attention to utilizing incineration technology to recover energy from waste (Waste to Energy Technology) because a large part of municipal solid waste is biomass waste.

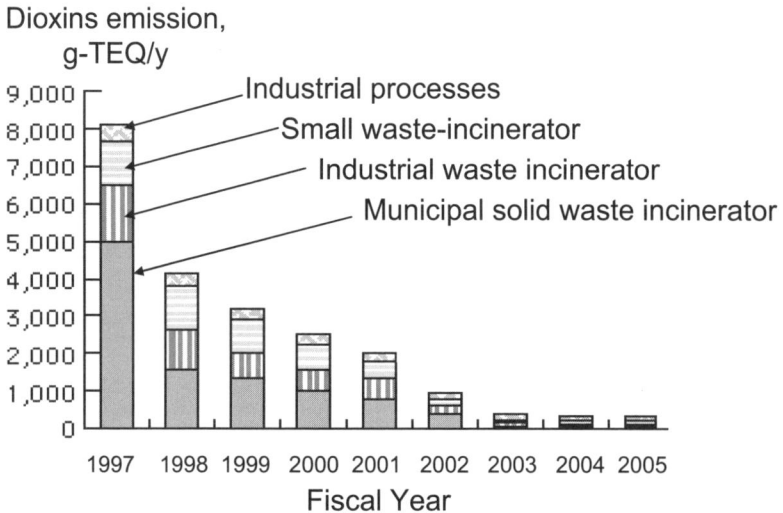

Fig. 5.4. Decrease in dioxins emission from all aver Japan (adopted and modified from the white paper, Ministry of Environment, 2007)

The case of illegal dumping on Teshima Island, Kagawa Prefecture shows a lesson of how costly it is to rehabilitate a damaged area once it is polluted. In 1978, a local company in Teshima got permission from the Kagawa prefectural government to operate as a disposal business for industrial waste like sludge from paper mill factories and food processing waste, waste wood chips, and animal manure, which was to be used to make a recycled product for the cultivation of earthworms. Nevertheless, the company had illegally dumped hazardous ASR (automobile shredder dust), waste oil and sludge to its disposal site on the island from 1983 to 1990, which caused serious and fatal pollution. The amount of the illegal waste, containing lead, chromium, cadmium, PCBs (polychlorinated biphenyls), dioxins, and so on, was estimated about 560,000 m^3. This illegal action was disclosed in 1990. The restoration cost was estimated about 45 billion yen, which was far beyond the capacity of the company to pay for without the help of public sector.

PCB waste treatment is another example of how difficult it can be to get public acceptance for waste disposal. After the ban on PCB and its production in 1974, the treatment of PCB waste was tried to be introduced, however, it could not get public acceptance mainly due to people's suspicions about secondary pollution through the exhaustion gas from the treatment facilities. Since then, a huge amount of PCBs have been just stocked and untreated for decades. The longer storage of PCBs creates a higher risk of unintentional spread of the PCBs into the environment due to the deterioration of the storage conditions, accidental leakage and unexpected accidents. Public awareness of this negative legacy of PCB gave rise instead to total opposition against PCB waste treatment facilities.

The Law Concerning Special Measures Against PCB Waste was established in 2001 and five PCB waste treatment plants have been constructed since then, including the earliest plant in Kita-Kyushu, which started operation in 2004.

5.2.3 Legislative Scheme of Waste Management and 3R

Fig.5.5 summarizes the current legislative scheme for establishing a Sound Material-Cycle Society, in which proper waste management and the promotion of the 3Rs are essential.

The "Fundamental plan on the environment" set by the Cabinet council determines the master plan for achieving the Low Carbon Society, and is based on the "Fundamental Law on the Environment." Under this master plan, the fundamental plan for establishing a Sound Material-Cycle Society also set by the Cabinet council based on the corresponding law, aims to

| Mater plan for the Low Carbon Society |
| Fundamental Plan on the Environment/Fundamental Law on the Environment |

Fundamental Plan for Establishing a Sound Material-Cycle Society		
Fundamental Law for Establishing a Sound Material-Cycle Society		
Objectives	to ensure material recycling in society
	to reduce consumption of natural resources
	to reduce environmental burden

Proper waste management/ Waste Management Law

Definition of waste: Principle of waste management: Establishment of standards for waste management: Regulations on waste collection/treatment/disposal businesses: Regulations on waste treatment/disposal facilities: Measures to control improper disposal: Public sector participation to develop proper facilities

Comprehensive scheme for promotion of 3Rs / Law for Promotion of Effective Utilization of Resources

Reduction and recycling of by-products: Product designing and manufacturing with awareness of 3R: Product labelling for separate collection of waste: Voluntary take-back and recycling of used products: Utilization of recyclable and reusable resources: Promotion of effective utilization of by-products

Regulation & Promotion of 3Rs for specific products/goods/wastes

Containers and Packaging Recycling Law	Home Appliance Recycling Law	Food Recycling Law
Consumers selectively discharge containers & packaging waste: Municipalities separately collect them: Businesses concerned recycle them.	Consumers bear collection and recycling cost: Retailers take back used home appliances from consumers: Manufacturers recycle them.	Businesses that manufacture, process and sell food products take charge of recycling food waste.
Construction Waste Recycling Law	End-of-Life Vehicle (ELV) Recycling Law	Green Purchasing Law
Construction contractors take charge of sorting dismantled construction materials and recycling them.	Car owners bear recycling cost: Manufacturers take back and recycle fluorocarbons, air bags and automobile shredder residue (ASR): Related businesses take back and provide ELVs.	National and local governments take initiatives in environmentally friendly products and services by stipulating their priority as specific procurement items.

Fig. 5. 5. Current legislative scheme of waste management and 3Rs (as of 2006)

ensure material recycling, reduce consumption of natural resources and reduce environmental burden. Circulative resources, including recyclable
waste, are defined and the concept of EPR (Extended Producer Responsibility), i.e. an environmental policy that extends a producer's responsibility
for a product to the post-consumer stage of the product's life cycle, is
manifested in this law.

Two major parts of the fundamental laws are the Waste Management
Law and the Law for Promotion of Effective Utilization of Resources,
which highlight the comprehensive policies and practices for proper waste
management and 3R implementation, respectively. Voluntary effort by
businesses is also important to ensure them. Voluntary guidelines for waste
treatment and recycling have been set by various industries. DfE (Design
for Environment) is also strengthening the implementation of the 3Rs. A
number of the guidelines of product assessment incorporating DfE are increasing. It is noted that CSR (Corporate Social Responsibility) is a key to
establish corporate governance over waste management and related recycling activities. This also reduces the potential cost of future environmental
restoration by avoiding the occurrences of unintentional illegal dumping
and improper recycling.

Furthermore, a series of laws have been enacted to regulate and promote
the 3Rs for specific products, goods, and wastes. The Container and Packaging Recycling Law, enforced in 2000 and amended in 2006, aims at reducing domestic waste by separately collecting waste containers and packaging, which accounts for about 60% in volume of the domestic waste.
The Home Appliance Recycling Law, enforced in 2001, aims at promoting
the proper recycling of television sets (cathode-ray tube type only), air
conditioners, refrigerators, and washing machines, all of which are bulky
and potentially hazardous.

The Food Recycling Law, enforced in 2001, aims to reduce and recycle
food waste generated at food related industries, such as food manufacturing, food wholesaling, food retailing, and food service. The Construction
Material Recycling Law, enforced in 2002, aims to promote sorting and
recycling of dismantled construction waste, such as concrete mass, wood,
and asphalt concrete mass. The End-of-Life Vehicle Recycling Law, enforced in 2005, aims to promote the recycling and proper disposal of end-
of-life vehicles (ELVs) by collecting and recycling air bags and ASR and
by collecting and then destroying CFCs (Chlorofluorocarbons) from ELVs.
The Green Purchasing Law, enforced in 2001, aims to ensure that national
and local governments take the initiative in purchasing environmentally
friendly products and services by stipulating their priority as specific pro-
curement items.

This legislative scheme has succeeded in establishing the comprehensive implementation of 3R activities and proper waste management.

5.3 Sound Material-Cycle - Domestic or International?

5.3.1 Improper disposal of solid waste in developing countries

Proper waste management and 3R activities are pursued in both developed and developing countries. However, differences can be found in the methods of their implementation reflecting the situation of developing countries, where it must be taken into the consideration the improper solid waste disposal and the health, welfare and recycling economy of the waste-pickers surviving on unsanitary landfill sites.

The implementation of proper disposal methods is most important. The step-by-step introduction of proper management technologies at disposal sites is recommended. As in developing countries, the least economically costly methods of solid waste disposal are still practiced. There is a need for gradual and economical improvement in order to close open dumps and create sanitary landfills. Table 5.2 provides an idea of the necessary steps towards the creation of sanitary landfills, such as control facilities, leachate purification, daily soil cover, gas collection (methane), buffer zones and measurement of the incoming waste (in order to estimate the landfill capacity left).

5.3.2 Transboundary movement of circulative resources

As the meaning of waste or circulative resources is relative, home appliance waste and circulative resources can be used to highlight the issue of the transboundary movement of waste and circulative resources. Japanese home appliances are exported to other countries as second-hand alternatives in some local markets outside of Japan. In fact, a significant amount of discarded home appliances was exported from Japan as shown in Fig. 5.6. It is noted that about 46% of the discarded television sets were exported as second-hand products or usable materials in FY2005. In the case of TVs made with cathode-ray tubes, which contain a large amount of lead, this means the used television sets exported are potential hazardous waste when they are discarded again after their second life.

These products' life is relatively short and they are mostly disposed of into open dumps without proper treatment. There must be the responsibility of producers in "taking care of the disposal of the hazardous components contained in the waste." On the other hand, the reuse of second-hand home appliances is not only economically preferable but also environmentally friendly in terms of the effective use of limited resources. This creates a controversial argument about whether a sound material-cycle system must lie in domestic or international.

Table 5.2 Necessary components in the step-by-step improvement to sanitary landfill (cited from a JICA text (2005))

Operation & Facilities	Necessary components in the step-by-step improvement of the level			
	Level 1	Level 2	Level 3	Level 4
Control facilities	○	○	○	○
Measurement of incoming waste	○	○	○	○
Landfill equipment	○	○	○	○
Approach and on-site roads	○	○	○	○
Enclosing bunds		○	○	○
Buffer zones		○	○	○
Daily soil covering & gas venting		○	○	○
Leachate circulation treatment			○	○
Mobile fence for litter prevention			○	○
Leachate purification				○
Seepage control works				○

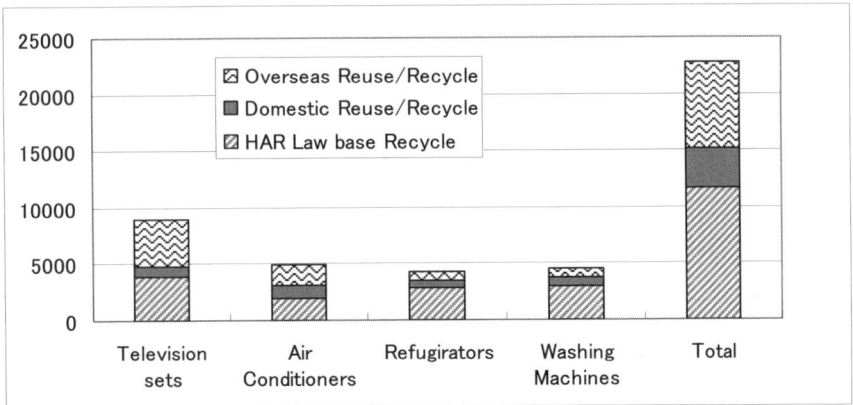

Fig. 5.6. Destination of discarded home appliances in FY2005 (Data source: documents, Assessment of Home Appliance Recycling (HAR) System division of the Central Environmental Council (2007))

Another argument is raised about whether exporting circulating resources that damage the domestic recycling industry is acceptable or not. This has happened in the case of PET (Polyethylene terephthalate) bottle recycling in Japan. The Japanese government has promoted the development of PET-to-PET bottle recycling by relevant private businesses with the high cost associated to technology and infrastructure development. In spite of this, a significant portion of PET bottle recycling has shifted to China, due to the economic benefits for municipalities that faced the decision of paying the high costs associated with domestic recycling or selling the PET bottles to be exported to China. This damaged the bottle-to-bottle recycling businesses in Japan. From a different viewpoint, PET bottle recycling costs through the Japan Container and Packaging Recycling Association have been drastically reduced, and were even reversed in 2006, which means the recycler paid money to get PET bottles, as shown in Fig.5.7. It is apparent that overseas market pressure led to a significant cost reduction in the domestic market.

The cost of the plastics other than PET bottles remains as high as 80,000 yen /ton (Fig.5.7). A similar situation could happen if the demand of waste plastic overseas is drastically increased in the near future.

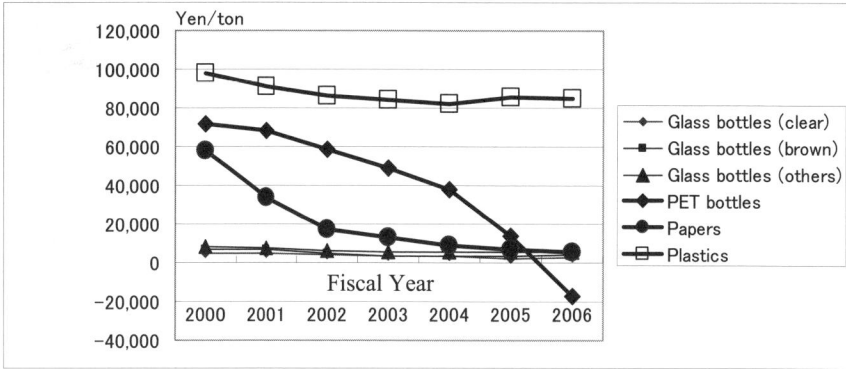

Fig. 5.7. Recycling cost of containers and packaging waste through the Japan Container and Packaging Recycling Association (JCPRA) (data source: a document of JCPRA, 2006)

References

[1] Clean Japan Center (2007) Waste management & 3R polices.
[2] Division of Assessment of Home Appliance Recycling (HAR) System, the Central Environmental Council (2007) Estimated flow of used home appliances.
[3] Institute for International Cooperation, Japan International Cooperation Agency (JICA) (2005) Supporting capacity development in solid waste management in developing countries, Chapter 2.
[4] Japan Container and Packaging Recycling Association (JCPRA) (2006) JCPRA News, No.34, August 2006 (in Japanese).
[5] Ministry of the Environment, Japan (2005) Japan's experience in promotion of the 3Rs, for the establishment of a sound material-cycle society.
[6] Ministry of the Environment, Japan (2006) Sweeping policy reform towards a "sound material-cycle society" starting from Japan and spreading over the entire globe: the "3R" loop connecting Japan and with other countries.
[7] Ministry of the Environment, Japan (2007) White paper on environment & sound material-cycle society.
[8] Mizoiri (2007), Solid Waste Countermeasures in Meiji Japan, Recycle-Bunkasha (in Japanese).

6. Dismantlement of Buildings and Recycling of Construction Waste

Tsuyoshi Seike

Noriko Akita

Graduate School of Frontier Sciences, The University of Tokyo
5-1-5, Kashiwanoha, Kashiwa, Chiba, 277-8563, Japan

6.1 Introduction

When a building is constructed, repaired or dismantled, building materials are consumed and waste is generated. The construction stage is divided into the manufacturing stage of the building materials at each factory and the composition stage of the building materials at the construction site. The manufacturing stage has achieved some success in reducing and recycling waste through the efforts of individual factories. On the other hand, during the composition stage, the packaging and broken pieces of building materials have usually been abandoned. Recently, major construction companies and house builders have begun to work on reducing and recycling this waste.

The efforts of "effective utilization of resources" in these manufacturing stages create a cost benefit, so it is advantageous for the factories and those who use the building materials. Therefore, to some degree, it is expected that effective utilization of resources in the manufacturing stage will continue to progress in the future.

On the other hand, at the repair stage, considering effective utilization of resources is still uncommon. Because the amount of resources and waste is so small, it has not been seriously considered. However, effective utilization of resources in the repair stage is sure to become more important because it is thought that the number of the buildings that last a long time will increase in the future.

Above all, effective utilization of resources at the dismantlement stage is the most important. 50 tons of waste or more are produced from the dismantlement of one detached house and several thousand tons of waste are produced from the dismantlement of one building. Because a large quantity of waste is generated in the building dismantlement stage, it is important to promote effective utilization of resources by recycling and re-using waste produced by dismantlement.

6.2 Japanese Housing Conditions

6.2.1 Short service life of housing stock in Japan

The housing stock in Japan has a short life compared with that in other countries. Fig. 6.1 indicates the service life of Japanese housing stock compared with other countries. In 1998, the average service life of housing in Japan was 42 years, while in the US and France the averages were both around 80 years. In the United Kingdom, the average was over 100 years. Thus, the service life of housing stock in Japan is very short compared with other countries. This condition causes a variety of serious problems in Japan.

Although the service life of the housing stock in Japan in 1993 was about 30 years, it was estimated to be 50 years in 2003. The service life of housing stock in Japan is still short, but is increasing gradually. According to the depreciation assets used for taxes in Japan, the life of reinforced concrete (RC) or steel reinforced concrete (SRC) structures is calculated as 47 years (statutory useful life). The straight-line depreciation method is specified for the calculation of a house's depreciation. The payment period by Japan housing Finance Agency for housing loans used to the fireproof building apartment owners amounts to a maximum of 35 years.

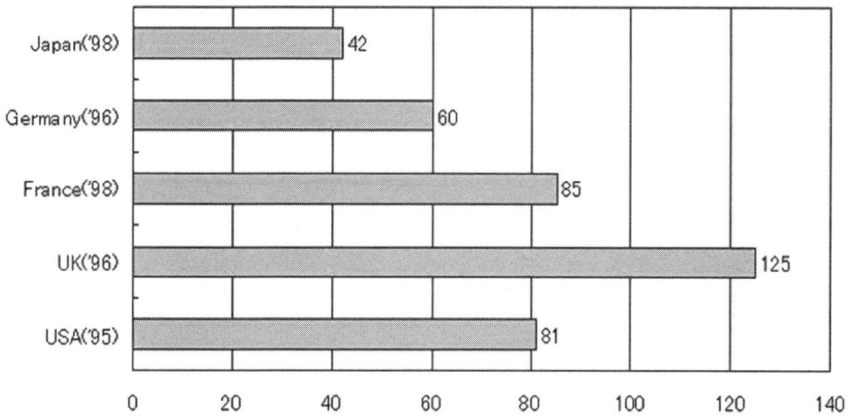

Fig. 6.1 Service life of housing in selected countries [1]

6.2.2 Housing construction boom in Japan

After World War II, Japan faced a serious housing stock shortage. The Japanese Government estimated that 4.2 million houses were urgently needed. This situation was solved quickly by a rapid increase in the rate of house construction. Fig. 6.2 shows the total number of housing units exceeded the number of all households in 1968, and by 2003 there were 1.14 times as many houses as households. In 2003, the total of all kinds of housing units in Japan exceeded 50 million. Fig. 6.3 also shows that over 60% of the buildings have been constructed after 1980.

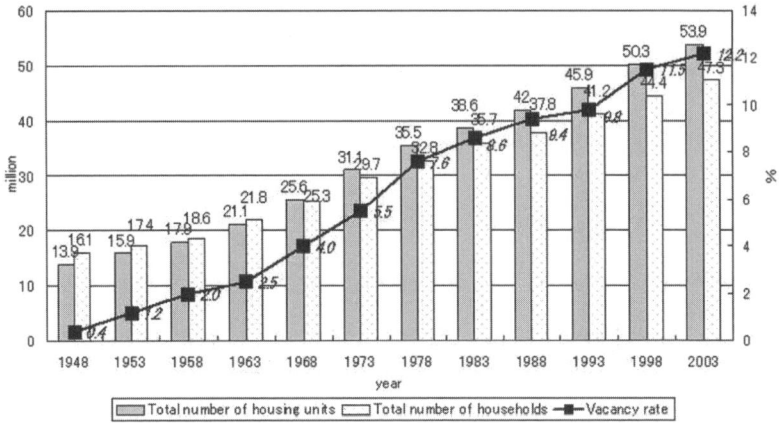

Fig. 6.2 Changes in the number of housing units and households [1]

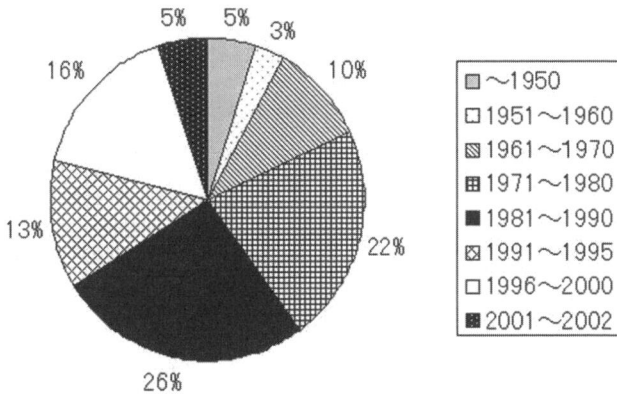

Fig. 6.3 Number of housing units based on the year of completion [1]

6.2.3 Types of housing in Japan

Fig. 6.4 shows types of housing such as detached houses, tenement houses and apartment buildings. In 1963, 70% or more of all housing units in Japan were single-family detached houses. However, the proportion of detached houses in Japan decreased gradually, and dropped to 56.5% of the total by 2003. On the other hand, the ratio of apartment buildings increased gradually and it reached about 40% in 2003.

In Japan, the housing type is related to the location. Urban areas tend to have more apartment buildings and fewer detached houses than rural areas. Fig. 6.5 shows that detached houses built in 2003 comprised about half of the total number of residential units.

The number of apartment buildings has increased gradually since 1963, and the ratio of detached houses is expected to decrease from 45% to 40% in Japan by 2007.

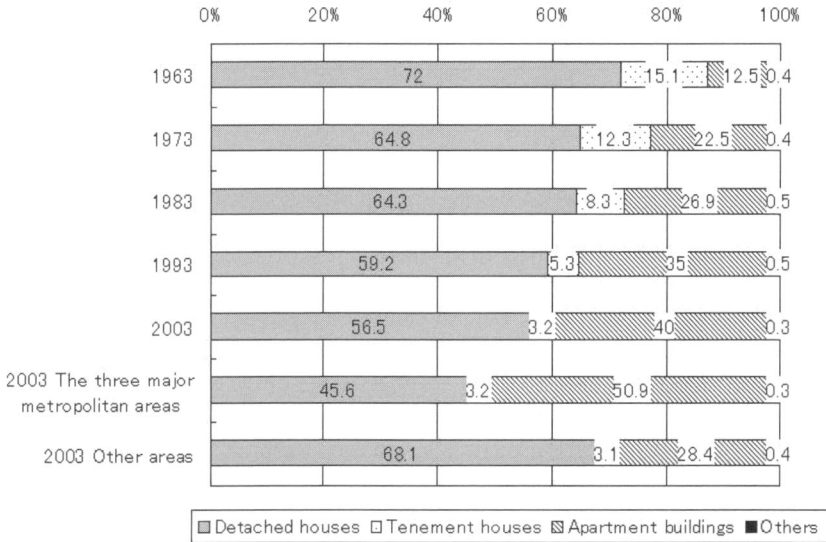

Fig. 6.4 Changes in the ratio of different types of housing units [1]

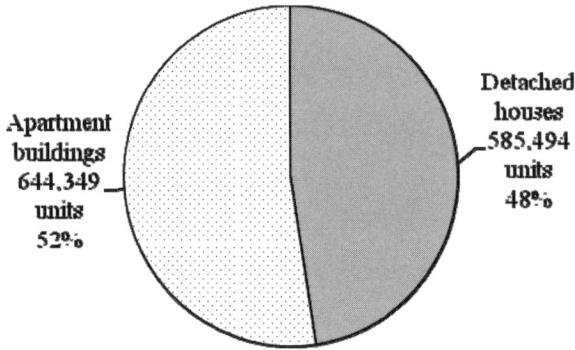

Fig. 6.5 Ratio of housing types in Japan in 2003 [1]

6.2.4 Structure of housing in Japan

The main construction type of detached houses in Japan is wooden. Wooden detached houses of both non-fireproof and fireproof structures made up over 60% of the housing stock in 2003 (Fig. 6.6). Wooden structures occupy about two-thirds of all the housing stock when classified by structure, but the proportion of non-wooden structures, such as RC houses and steel-frame houses, is increasing. The ratio of non-fireproof wooden houses to fireproof wooden houses is also increasing. Fireproof wooden houses are wooden buildings with roofs and outer walls coated with fire-preventive materials, such as mortar or zinc-coated sheet iron. Steady progress is being made towards higher rates of non-flammable or fire-resistant wooden houses. Compared with rural areas, urban areas have more non-wooden and fireproof houses.

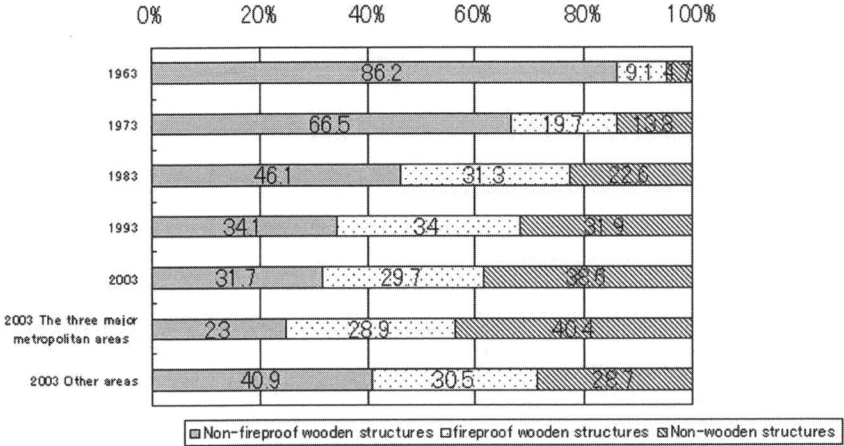

Fig. 6.6 Changes in the percentages of types of wooden housing stock [1]

6.3 Status of Construction Waste in Japan

6.3.1 Current condition of construction waste in Japan

Waste derived from construction comprises about 20% of the total amount of industrial waste in Japan. Fig. 6.7 and Fig. 6.8 shows the exhaust ratio of industrial waste in Japan by industry and by type. Although construction waste is also generated from civil engineering projects such as roads and bridges, most construction waste is generated from the dismantlement of buildings. Therefore, it is necessary to control the process of dismantlement and promote the recycling of dismantlement waste. Fig. 6.9 shows the change in the amount of exhaust waste from construction according to fiscal year in Japan. The quantity of construction waste decreased from 100 million tons in 1995, to 82 million tons in 2002 and to 77 million tons in 2005, owing partly to the reduction of public utility works and decreased housing construction starts. But the main reason for the decreasing amount of waste was the Construction Material Recycling Law of 2000. The recycling rates increased from 58% in 1995 to 92% in 2002 and 92% in 2005. The quantity of the final disposal of the waste was reduced from about 42% (42 million tons) in 1995, to 8% (7 million tons) in 2002 and to 8% (6 million tons) in 2005.

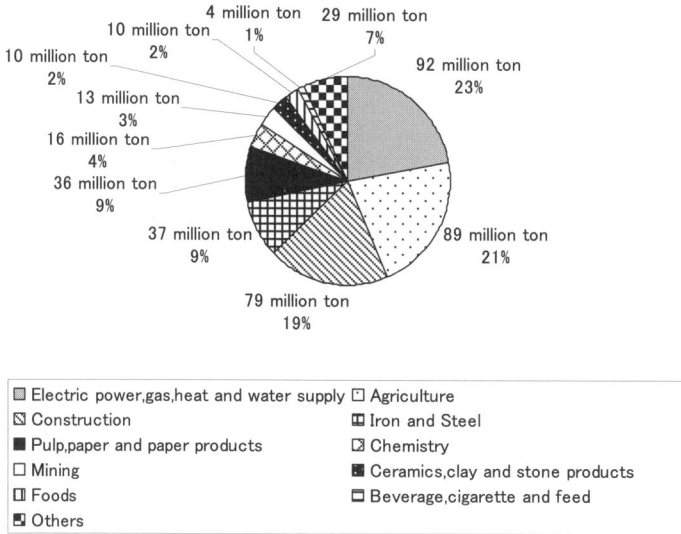

Fig. 6.7 Quantity of industrial wastes by industry [2]

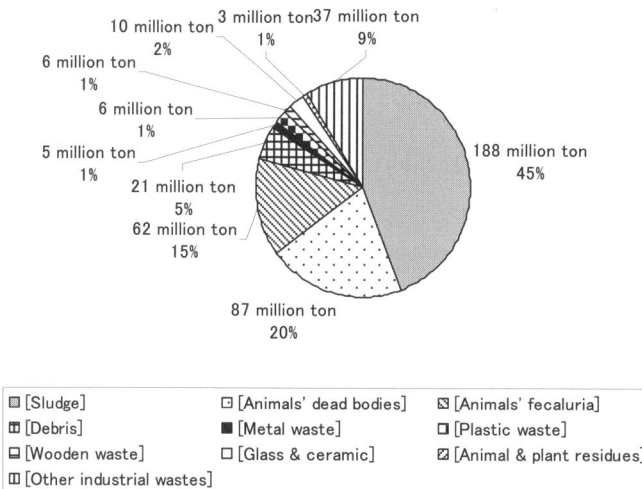

Fig.6.8 Quantity of industrial wastes by type [2]

National Sum Total
100 million ton(1995)

National Sum Total
82 million ton(2002)

National Sum Total
77 million ton(2005)

☐ demolition   ☐ new construction work
☒ public civil work   ⊞ private civil work

☒ Recycling   ▤ Final Dumping
▨ architecture   ☐ civil work

▨ architecture ☐ civil work

(Unit = million ton)

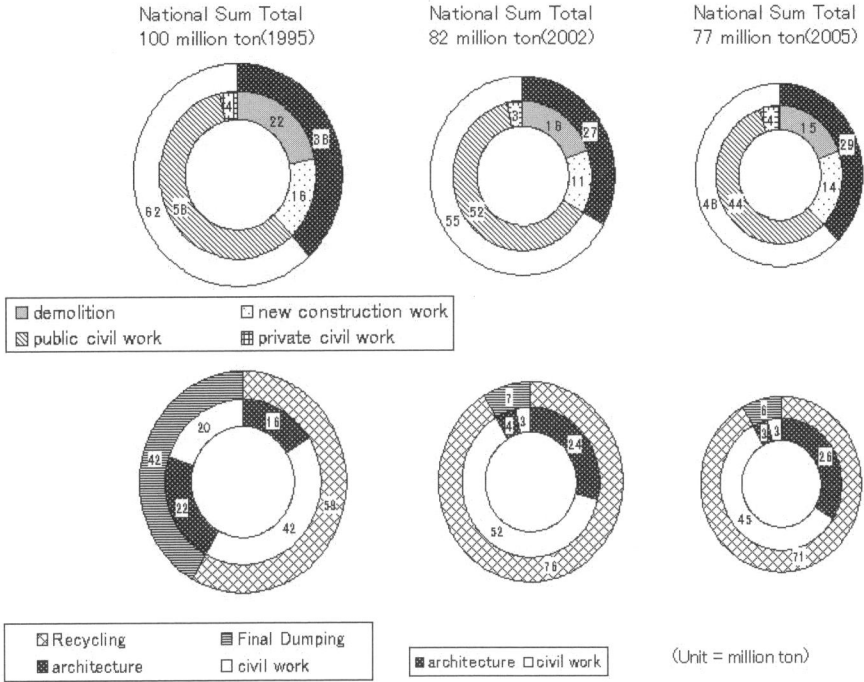

Fig.6.9 Final dumping and recycling rate of construction wastes [3]

6.3.2 Waste generated in the dismantlement process

When a house is dismantled, a huge amount of waste is generated. Fig. 6.10 shows the waste ratio and total waste amount for a RC apartment building. Most of the waste is concrete. However, with detached houses, half of the waste is concrete used for the foundation and basement, and wooden waste is only 20% of the total (Fig. 6.11). A variety of mixed waste such as glass, plastic, and ceramics are generated from house dismantlement. The mixed waste makes various problems of dismantlement and recycling. But the problem is not so severe for RC apartments because the waste is mostly concrete. However, in the case of detached wooden houses, the separation of waste is difficult because the types of waste are varied and complex. Even if the waste is impossible to recycle, it is necessary to separate all waste generated by dismantlement. Separation must be done onsite.

Plastic waste, 0.7ton, 0%

Glass and Ceramics waste, 0.4ton, 0%

Wooden waste, 0.8ton, 1%

Inorganic compound with cement, 0.04ton, 0%

Mixed waste, 4ton, 3%

Metal waste, 8ton, 5%

Debris of Concrete and Mortar, 140ton, 91%

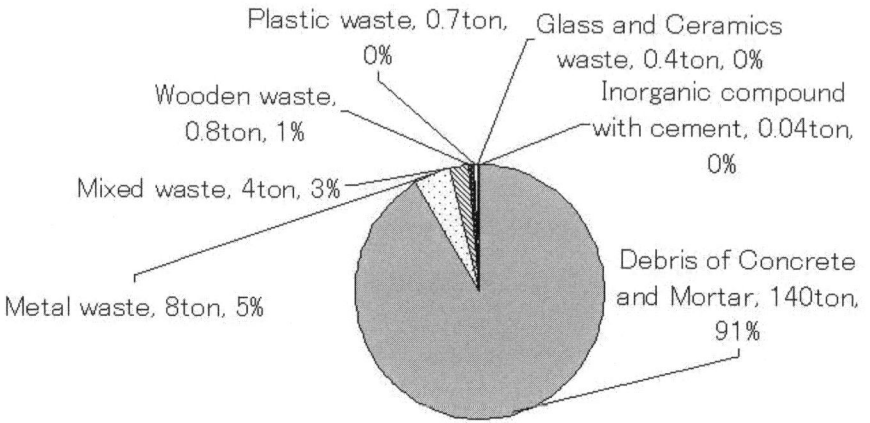

Fig.6.10 Waste output from a RC apartment building [4]

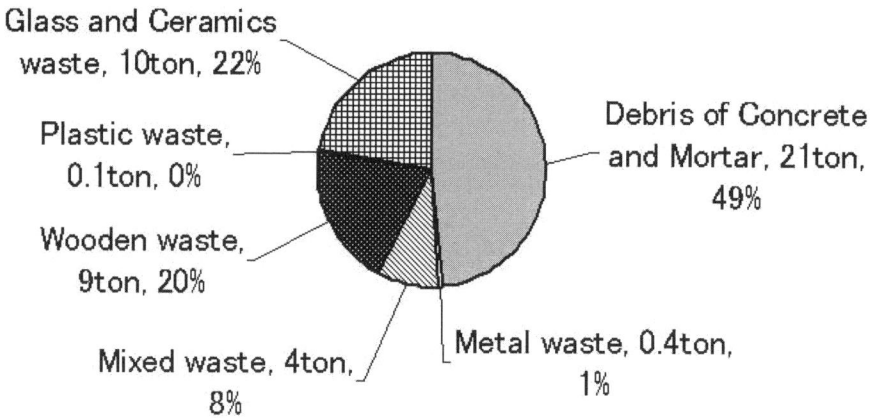

Glass and Ceramics waste, 10ton, 22%

Plastic waste, 0.1ton, 0%

Wooden waste, 9ton, 20%

Debris of Concrete and Mortar, 21ton, 49%

Mixed waste, 4ton, 8%

Metal waste, 0.4ton, 1%

Fig.6.11 Waste output from a wooden detached house [4]

6.4 Restriction of Waste Disposal in Japan

6.4.1 Construction Material Recycling Law

The process of dismantlement has been subject to governmental oversight since the Construction Material Recycling Law of 2000. The situation of dismantlement waste in Japan has changed since this law was enacted. The law imposed the obligation of appropriate dismantlement and recycling. Moreover, contracts concerning dismantlement procedures must be exchanged between the owner and the dismantlement company of the building. Dismantlement companies have to follow appropriate procedures (Fig. 6.12). At the same time, a system for registering a dismantlement company was established by the law. However, we have to recognize that improper dismantlement and illegal disposal is still being committed. Although there is the obligation to obey the law, recycled construction waste is only 70% of what should be recycled.

The techniques of dismantlement for a large-sized building and a small-sized house differ widely. The method of dismantlement for a large-sized building is mostly established, but for a detached small-sized house, it is still under examination and not established yet. We have to establish a method of dismantlement for the small-size houses that make up half of the total number of houses in Japan.

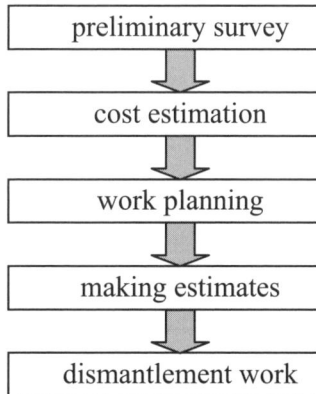

```
┌─────────────────────┐
│  preliminary survey │
└─────────────────────┘
          ↓
┌─────────────────────┐
│   cost estimation   │
└─────────────────────┘
          ↓
┌─────────────────────┐
│    work planning    │
└─────────────────────┘
          ↓
┌─────────────────────┐
│   making estimates  │
└─────────────────────┘
          ↓
┌─────────────────────┐
│  dismantlement work │
└─────────────────────┘
```

Fig. 6.12 Dismantlement flow chart

6.4.2 Proper procedure for dismantlement

Before the dismantlement construction starts, a preliminary survey is need that estimates the amount of materials, hazardous materials, etc. The next step is cost estimation. The cost includes making a temporary building cover preventing trouble to neighbourhoods and scaffolding for the dismantle work, all kinds of dismantlement works, waste collection, transportation and so on. The dismantlement work process is described in the "work plan." The work plan includes the dismantlement work process, the dismantlement technique and the material separation method. It also contains an explanation of the waste content to be seen by the former owner and the current owner. After this preparation period, making estimates and dismantlement works follow. Dismantlement works in Japan will be explained in the following section.

In Japan, there are three important points to consider in the establishment of a proper dismantlement method. One is the establishment of a proper technique for dismantlement, the second is to promote making proper work plan, and the third is the establishment of a proper cost. After 2000, owners of the buildings have had to pay a much higher cost for dismantlement by the Law.

6.5 Procedure for dismantlement

6.5.1 Detached house dismantlement in Japan

First, lighting fixtures, unit baths, kitchen cabinets, built-in air conditioning units, etc. are removed (Fig. 6.13). In general, the building materials consist of complex materials including wood, metals and plastics that are very troublesome to sort later on. The Construction Material Recycling Law makes it mandatory to dismantle building materials by hand (except when difficult for structural or technical reasons). However, in practice, building materials are often dismantled by machine together with the building frame, and they are then sorted without manual sorting.

Fig. 6.14 shows the removal of interior finishing materials like boards and ceilings. Fittings, interior wall materials, ceiling materials, floor materials, etc. are dismantled. In many cases, except for wooden housing, wooden interior finishing materials are separated before dismantling by machine, while metal materials are separated after machine dismantlement.

After removing the interior building materials, workers remove the roof tiles and roofing materials that can be separated for recycling from the top of the buildings. The picture on the left side of Fig. 6.15 shows the manual dismantlement of a modern thin roof. The Construction Material Recycling Act demands the removal of roofing materials by hand. But for some kinds of roofing materials, removal by machine allows for sorting so that it can be more efficient and safer. Therefore, removal by hand cannot always be said to be adequate. Since rainproof materials become a disturbance in the case of the reproduction of concrete, it is classified after machine dismantlement. The removed materials are then transported to the recycling yards. Exterior walls and superstructures are often dismantled by heavy machinery (Fig. 6.16). When the site is not wide enough to allow for heavy machinery, manual dismantlement is also employed. The right picture of Fig. 6.15 shows the dismantlement of a wooden house by hand. In the case of dismantling by machine, water is often sprinkled during dismantlement for dust prevention. Compared to the case of wooden houses, steel structure and RC houses tend to be dismantled by machine more often. There are two different ways of recycling steel structures and RC structures. At the present, autoclaved lightweight concrete (ALC) waste used for steel structures is not reused, just separated; RC waste is reused as a road-building material. The dismantlement method for steel and RC houses is selected according to the height and structural strength of the building, spatial capacity of the site, etc. Buildings constructed with RC are usually dismantled by a concrete crusher. On the other hand, buildings constructed with steel are dismantled either by a steel cutter attached to heavy machines, or manual gas cutting.

The foundation is dismantled by heavy machinery. Concrete lumps are further divided by separate reinforcing bars, which are then carried out as needed. Since the majority of waste from the foundation portion consists of concrete, it is decomposed after demolition and carried to a recycling yard. The method for dismantlement by heavy machine for high-rise building is to begin on the highest story and descend, or to install the machine on the ground, depending on the dismantlement conditions.

The last step is to sort the waste and dispose of it. In general, adequate space for sorting various wastes is not available at the dismantlement sites of small houses, but the dismantlement sites of apartment buildings do have enough space as a whole. In Japan, this is one of the problems at the dismantlement sites because there are a lot of detached houses built on small sites. In the case of small site dismantlement, the dismantlement wastes are stored in a parking space, or outside space, if available, or inside of the dismantled building and put the containers. When there is no storage place, dismantlement wastes are carried directly to a truck from each room. In this case, dismantlement wastes are temporarily placed inside the dismantled building. But, this lowers work efficiency and increases the degree of mixed dismantled wastes. Moreover, this method may be unsafe for workers.

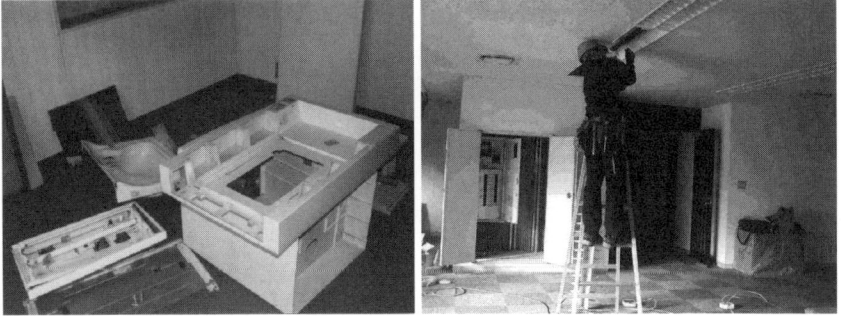

Fig. 6.13 Equipment removal process

Fig. 6.14 Interior finishing removal process

Fig. 6.15 Dismantlement by hand

Fig. 6.16 Dismantlement by machine

6.5.2 Cases of building dismantlement in foreign countries

The dismantling method in foreign countries is influenced by the circumstances of recycling and the disposal method of building waste in each country. Because a common, strict regulation regarding waste was adopted in the EU, each country should be taking new action for dismantlement. But in practice, some actions are different in each country and some actions are common.

Generally, there are no large differences between the fundamental technologies for dismantling buildings between countries. It is common that interior decorations are dismantled by hand, but sometimes a small machine is used. Of course, a heavy industrial machine is used for dismantling the skeleton of the building. There are small differences in the types of tools used for dismantlement from country to country, but the tools are similar. Waste is transported to the waste disposal spot and treatment facility after being separated into wood and glass, which is produced mainly by interior decorations, and concrete and iron, which is produced mainly by the skeleton. The dismantlement in other countries does not use particularly superior technology in comparison with Japan. However, the methods have aspects that are widely different from Japan, and there are also different aspects according to countries.

Germany

There is the company which handles mainly a feasibility study of the dismantling construction in Germany. Because buildings in the EU have a long service life, repairs are generally added to the building many times. For example, in the case of a 100 year-old building, it may be composed of pre-war materials, materials from repairs in the 1960's and materials from repairs to the exterior wall made recently. Therefore, at the time of the dismantlement, it is necessary to investigate at first what kinds of materials have been used.

Japanese buildings have a short service life, which makes it easier to identify the types of materials that have been used compared with EU countries. In a case of Germany, sections from the inner wall, the floor and the ceiling wcrc takcn off, a sample was gathered, and dismantlement began only after the safety of the material was confirmed. The reason why such a system functions adequately in Germany is that the company which handles mainly the investigation that mentioned above exists.

Fig. 6.17 and 6.18 show one example of a bank office building dismantlement in Germany. This building was repaired many times since its construction in 1910 until it was finally dismantled in 2002. These pictures show the inside of the building where an inner wall had been removed. Building materials containing asbestos are confined to a special bag marked by the letter "a." Nothing is left on the floor. Fluorescent lamps, papers and rubber are also separated. Fig. 6.19 shows a container for collecting material wastes and moving them to the recycle yards. These containers are carried in the early morning or at midnight. Fig. 6.20 shows investigators checking the floor materials to tell whether or not the materials are hazardous.

Fig.6.17 The separation of the inner wall and separation of the asbestos

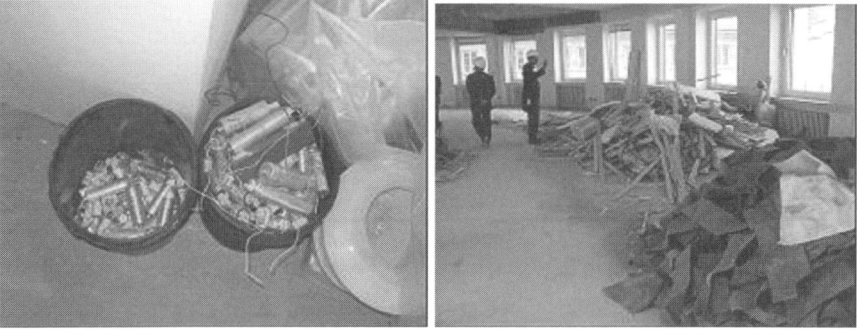

Fig. 6.18 Fluorescent lamps, paper and rubber are also separated

Fig. 6.19 Container for collecting material wastes

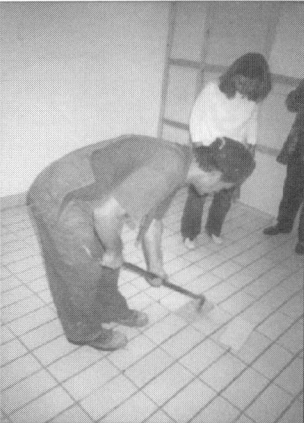

Fig. 6.20 Investigators checking the floor materials

France

In France, lead in paint used for interior decoration is becoming a social problem, and there is careful management of it at the time of the dismantlement. There are specialized companies for the analysis and removal of lead paint. The companies undertake feasibility studies beforehand and draw a "P" on the wall to indicate where lead paint has been used. When a "P" is drawn on a repair area, a special procedure to confine or remove the lead is needed. When dismantling the building, carefully removing the parts with lead is necessary.

Fig. 6.21 shows a dismantlement site in France. A "P" has been drawn on the frame after an inner wall has been removed. The proper management of the toxic substance is taken strictly in France, and the business market of that purpose exists. As for asbestos, very intense management of it is already carried out in the EU, unlike Japan, where it has suddenly become a severe problem recently. In the aspect of the separation of the toxic substance, it may be said that the management of the French dismantling construction is carried out strictly from Japan.

The regulation parameters and methods for dismantling are different by country. Lead paint is distinguished in France because it is dangerous, but similar efforts are not taken in Germany and the U.K. From various standards for dismantling, it is important to find a suitable method to be adopted by each country and for each case.

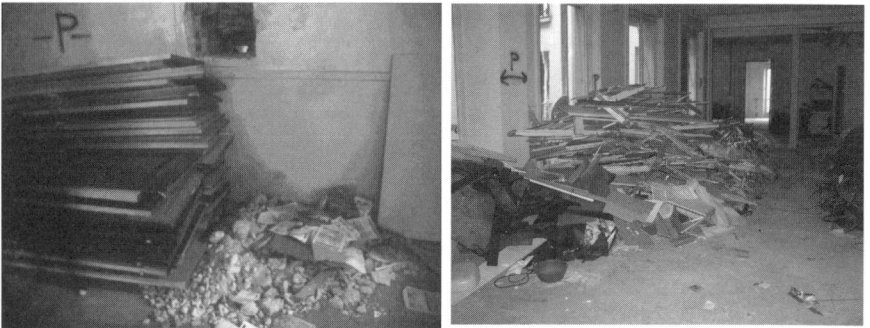

Fig. 6.21 A "P" on the wall indicates lead paint

United Kingdom

A process called "salvage" is undertaken at dismantling spots in United Kingdom (Fig. 6.22). This is the removal of stones from the decorations and old bricks from structures, when reuse is possible, before starting the main dismantling. The salvaged stones and old bricks are separated and brought to reuse markets or another construction site. Such an old materials are used for the new building to take the image of an old building over. It is also used for other buildings as old, tasteful building materials. Because old building materials, including old bricks, are regarded as valuable in the U.K., every year about 3 million tons of recycled building materials are provided from building dismantling spots. There is a market for reuse materials, salvage takes place at dismantlement spots in United Kingdom..

Fig. 6. 22 British dismantling spot and "salvage"

6.6 Recycle and reuse

Germany

Fig. 6.23 shows a waste treatment facility near Munich. Separation work by the hand is taken to separate the materials which are different from the material of the specific recycling purpose at the first stage. After separating wastes by hand, it will be separated by a machine. These steps are common all over the world.

In Japan, concrete and bricks are crushed and the pieces are used for road materials. But they are used for private gardens in Europe. Fig. 6.24 shows the specification of recycled inorganic materials which obtained in Germany It shows that using different qualities of recycled materials for different places, and that the public sector must use these recycled materials. But they usually buy the poorer quality materials.

Fig. 6.23 Waste disposal and treatment facility in Germany

Fig. 6.24 A specification that obtained in Germany

China

The case of reuse and recycling in China is a little different from EU countries. In Beijing, to gather soil and produce new bricks has been forbidden in order to protect farmland. The main problem in Beijing is the lack of building materials. So bricks are collected and reused from dismantlement sites. It is easy to find people who collect material resources in China. Fig. 6.25 shows a dismantlement site in Beijing. It shows that bricks are removed, and will be carried to a construction site or a used building materials market. Fig. 6.26 shows workers carrying building materials and a used building materials market in Beijing. Wooden materials such as beams and panels are also collected from dismantlement site and are sold markets. Fig. 6.27 shows used floorboards at the used building materials market. In Japan, we do not reuse these low-quality materials, but China reuses such wood actively for lack of raw materials.

Fig. 6.25 Dismantlement site in Beijing

Fig. 6.26 Gathered and carried used building materials sold at the market

Fig. 6.27 Floorboards on sale at a building materials market in Beijing

6.7 Summary

There are some regulations about the use of recycled materials in Japan. However, recycled materials are used in Japan only when raw materials are scarce. We have no proper market or demand for recycled materials yet. This is the primary problem in Japan. Therefore, to promote construction waste recycling in Japan, it is important to investigate the feasibility of a method and the cost of a recycle market.

The technologies for dismantlement and recycling have already been established, but a system for appropriate dismantlement and a recycling market has not matured yet. At the same time, it is important to offer ecological technologies to developing countries. Also, it is necessary for us to publicize good technologies of dismantling and recycling.

References

[1] Japanese Ministry of Internal Affairs and Communications: Housing and Land Survey (2003)

[2] Japanese Ministry of Environment: waste disposal treatment, discharge and handling of industrial waste survey (2004)

[3] Japanese Ministry of Land, Infrastructure and Transport: The Investigation into the actual conditions about by-products concerned with construction (2002)

[4] Data from Seike Laboratory and Building Research Institute and project research (2002).

7. Holistic Approaches for Impact Assessment in Urban Environmental Management

Toshiya Aramaki

Department of Urban Engineering, The University of Tokyo
7-3-1, Hongo, Bunkyo-ku, Tokyo, 113-8656, Japan

7.1 Introduction

In order to create a better urban environment, various measures which aim to reduce environmental burdens and mitigate specific environmental impacts should be conducted. However these measures often shift environmental burdens to other areas or other fields, or create new burdens to the environment. Therefore, we need to grasp various environmental impacts which may be caused by these measures holistically, and choose appropriate measures for sustainable management of the urban environment. This chapter introduces several tools and approaches for a holistic assessment of the environmental impact of urban environmental management.

7.2 Metabolic and Material Flow Analysis

Urban areas consume a wide variety of resources, such as air, water, energy, food and many other products to sustain their activities, then discharge them after use in the form of gas, waste heat, wastewater and solid waste. Due to the similarity to the processes of living organisms, this process has been called "Urban Metabolism." "Urban Metabolism" is a widely used concept in the field of environmental management, and it is now recognized that inappropriate management of this metabolism causes environmental problems in urban areas such as air, water and soil pollution, urban heat island and others. Appropriate management of this urban metabolic process would contribute not only to solving environmental problems inside urban areas, but also would reduce the amount of resource consumption and discharge to the outer environment, thus partially helping to solve regional and global environmental problems.

Metabolic Analysis is a tool to analyze what enters, what leaves and to what degree entering material is utilized usefully or is lost for a specific subject. Subjects of this analysis can be a factory or a business, but also a region or a nation [1]. Through this analysis, the opportunity to improve environmental performance for a target subject can often be identified.

Figure 7.1 shows the total material flow for Japan [2]. In this example, "Japan" was considered the subject of metabolic analysis, and the amount of input, output, consumed, and remained inside Japan were analyzed in terms of weight. Almost 40% of material used in Japan came from foreign countries, and almost half of this amount remained: mainly as buildings and infrastructure. 10% of material input was recycled to use as another resource, and 2.5% of material input was disposed of in landfill sites.

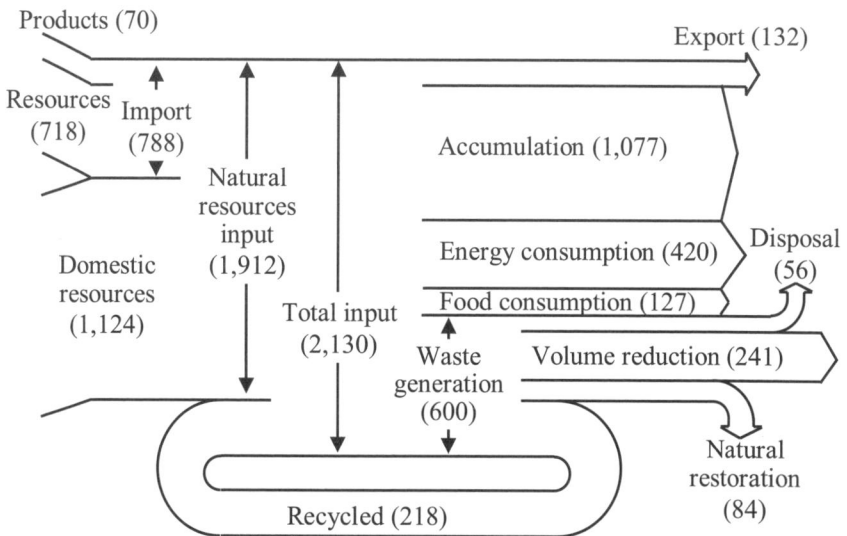

Fig.7.1. Material flow in Japan in 2000 (million-ton/year), translated and reproduced from the Ministry of Environment [2]

Through this figure, we can understand the current situation of material use, and may consider opportunities to improve it as part of the goal of achieving a more sustainable society. We can also analyze specific materials to understand the situation of each material precisely and to advance more concrete proposals for improving its flow.

When we focus on the flows not only across the boundaries of a subject, but also within a subject, then analyze these flows, it is called "Resource Flow Analysis" or "Material Flow Analysis (MFA)." Many studies to date have been focused on one element (atom) or one substance, such as metals and greenhouse gasses, as the target subject, then called "Elemental Flow Analysis" or "Substance Flow Analysis" [1]. "Material Flow Analysis" is an attractive tool for solid waste and wastewater management, with organic and nutrient flows considered as the target areas. There have been several studies to analyze organic and nutrient flows for discussing the influence of solid waste and wastewater management [3,4,5].

Figure 7.2 shows the estimated nitrogen flows in urban districts of Haiphong city, the third largest city in Vietnam [6]. In this analysis, the nitrogen flows related to solid waste and sanitation issues are focused, then eight components for nitrogen flows: agriculture, food factories, food markets, households, solid waste collection, septic tanks for human excreta treatment, organic waste composting and landfills are selected. It indicates that food imports account for two-thirds of total nitrogen input to the target area. Raw materials to industry and the import of chemical fertilizer account for more than 10% of the total nitrogen input. Regarding the nitrogen output from the target area, direct disposal of septage to the environment has the highest contribution with 20% of the total output. Discharge of seepage from septic tanks and grey water also made up a significant portion, around 15% of the total output. Food imports are hard to control in terms of nutrient management, but appropriate management of septage from septic tanks can be managed and is the most important issue as identified through this analysis.

7.3 Induced environmental burdens and Life Cycle Assessment

If we install a treatment plant for septage, and remove nitrogen by denitrification or by composting the product, we can decrease discharge to the environment by a significant amount, as shown in Figure 7.2. However, construction and operation of a treatment plant may require a huge amount of energy and materials, such as cement and steel. In the process of mining, refining and transporting these inputs, we may discharge pollutants into the environment. This means that we may shift environmental burdens from local nitrogen pollution to other regional or global problems. Such a

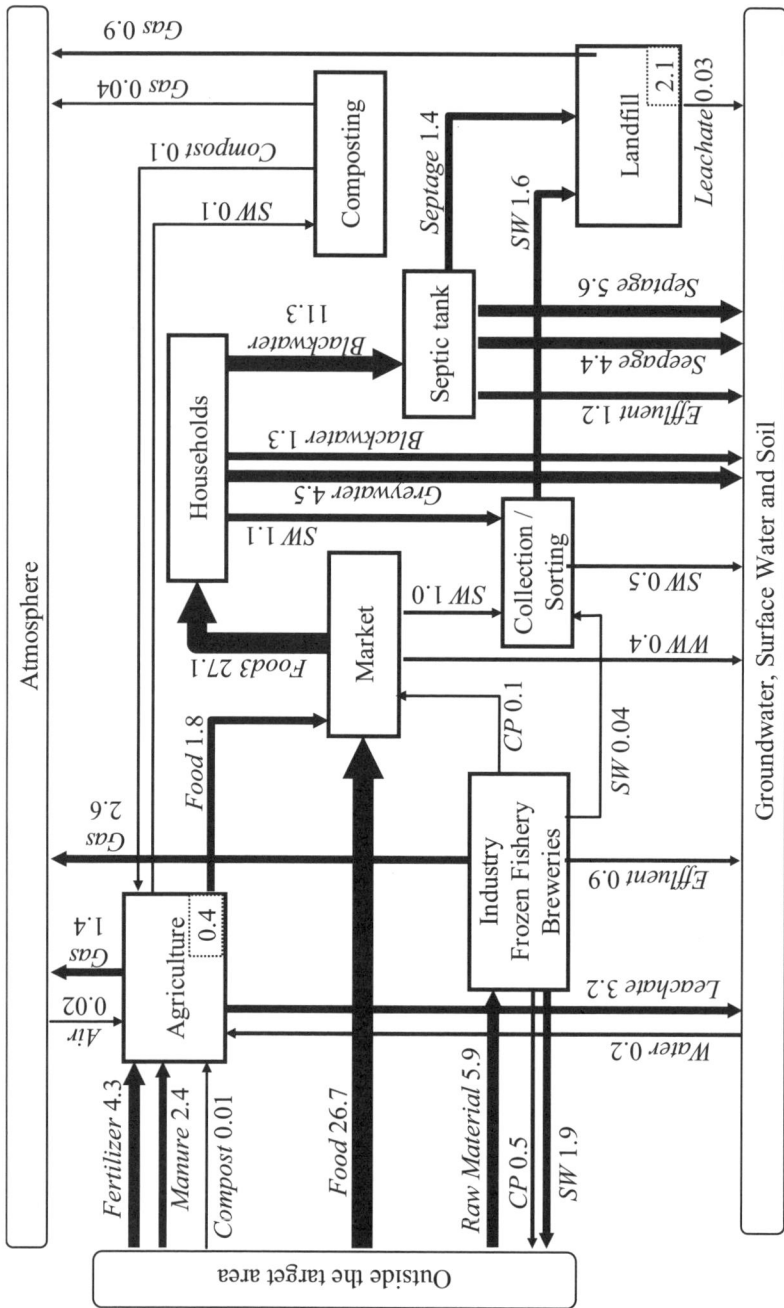

Fig.7.2. Nitrogen flows related to sanitation in urban districts of Haiphong City (*CP*: consumer products, *SW*: solid waste, g/person/day)[6]

induced environmental burdens are sometimes significant, so careful consideration is necessary when we plan to have such facilities.

Life Cycle Assessment (LCA) is a widely used tool for evaluating direct and induced environmental impacts associated with products and systems. The focus of LCA is on the entire life cycle of the product, i.e. from the extraction of the raw materials through the production of materials and components, then through the manufacture, transportation and use of the product and finally to the product's disposal or possible recycling, as shown in Figure 7.3. The methodology of LCA contains four main phases: (i) Goal and Scope Definition, (ii) Inventory Analysis, (iii) Impact Assessment, and (iv) Interpretation [7], as shown in Figure 7.4.

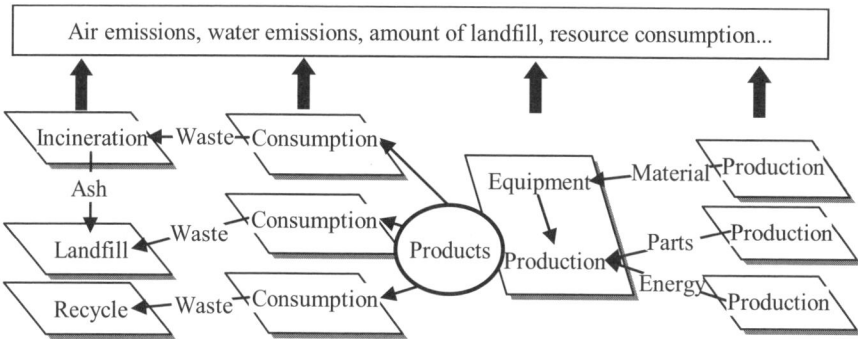

Fig.7.3. Life cycle of products and induced environmental burdens

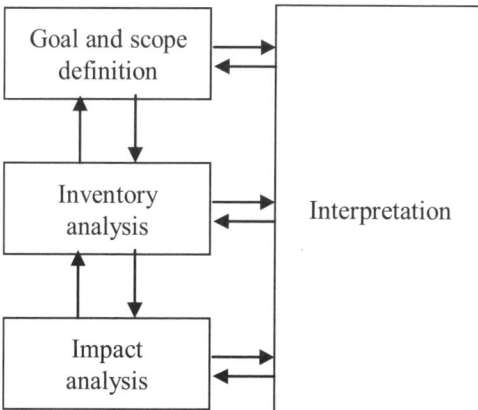

Fig.7.4. Process of Life Cycle Assessment

In the first phase, the goal and intended use of the LCA, the scope of the assessment concerning system boundaries, functions and flows, the required data quality, the technology and the assessment parameters are defined. Then, information on inputs (resources and intermediate products) and outputs (emissions, wastes) for all the processes in the target system is collected in the Life Cycle Inventory analysis (LCI). Life Cycle Impact Assessment (LCIA) is the phase where inventory data on inputs and outputs are translated into indicators about potential impacts on the environment, on human health, and on the availability of natural resources. Finally, the results of the LCI and LCIA are interpreted according to the goal of the study and sensitivity and uncertainty analyses are performed to qualify the results and the conclusions in the interpretation phase.

I would like to introduce one example of LCA application for municipal solid waste management [8]. Figure 7.5 shows the system boundary for the treatment systems of municipal solid waste in Sao Paulo for evaluating the environmental performance of direct landfill and incineration. It is assumed that they have a power generation facility in the incineration plant, and a leachate treatment facility in the landfill. Generated electricity in the incineration plant is assumed to be substituted for electricity from the power company, and power generation at the power company is also considered within the boundary. According to this boundary, the life cycle inventory for treating one ton of waste of average composition has been estimated in Table 7.1. Figure 7.6 shows the estimated impact on global warming, acidification and nutrient pollution by these two alternatives. These values are calculated from inventory results, and a set of impact potentials for each pollutant on target environmental issues such as Global Warming Potential (GWP) has been estimated.

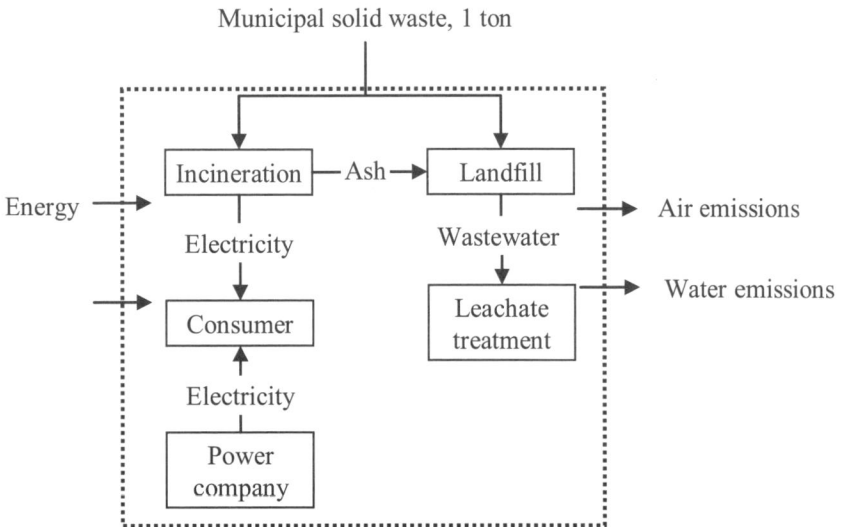

Fig.7.5. System boundary for LCA on municipal solid waste treatment

Table 7.1. Life cycle inventory for the treatment of 1t of municipal solid waste of average composition in Sao Paulo

	Incineration	Landfill
Inputs		
Energy		
Electricity (kWh)	123	665
Diesel (l)	0	1
Natural gas (m3)	8	0
Materials and ancillary substances		
Limestone (kg)	2	0
NH_3(kg)	1	0
NaOH(kg)	9	0
Outputs		
Emissions to air		
Carbon dioxide, fossil[a] (kg)	571	2
Hydrogen chloride (g)	6	6
Hydrogen fluoride (g)	0.3	1
Hydrogen sulfide (g)	0	18
Methane (g)	0	35,849
Nitrogen oxide (g)	367	41
Nitrous oxide (g)	100	2
Sulfur oxide (g)	6	32
Emissions to water		
Total nitrogen (g)	0	1003
Solid residues		
Fly ash (kg)	38	0
Recovered materials and energy		
Electricity (kWh)	564	0
Heat (MJ)	2,029	0

[a] Biogenic CO_2 was not included because it is not considered a contributor to GWP.

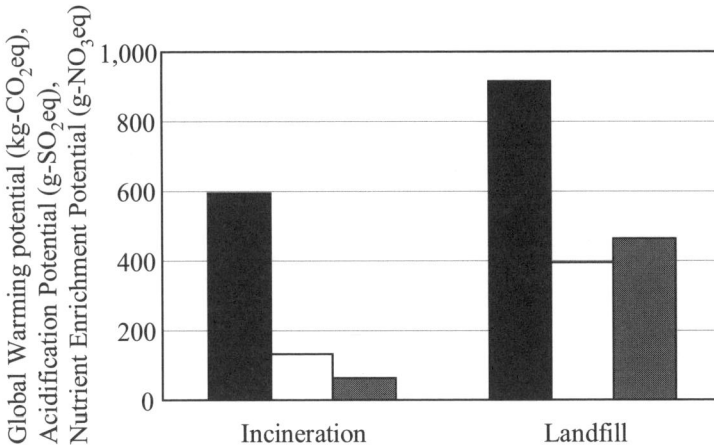

Fig.7.6. Life cycle impacts of 1 ton of municipal solid waste by incineration and landfill treatment (Left: GWP, Middle: AP, Right: NEP)

Through this analysis, we can recognize that direct landfill has a larger impact for all three categories. If we check the contribution of each pollutant or process to the overall environmental impact, we can discuss the opportunities and methods to mitigate these environmental impacts. For example, a main contributor to GWP in landfills is methane generation during biological degradation of organic waste, so avoiding methane emissions is important in landfill management. In the case of incineration, the GWP is largely due to burning plastics, which create anthropogenic CO_2 emission. Producing electricity in incineration plants did not decrease the GWP considerably because electricity in the grid is mainly hydro-generated in the target area.

7.4 Comparison of different environmental aspects

In order to choose better alternatives for environmental management, we need to compare the positive and negative impacts caused by direct environmental burdens, as well as the potential impacts caused by induced environmental burdens shown in LCA. In many cases, however, environmental impacts caused by direct burdens are different from those by induced burdens. In some cases, a specific environmental burden influences several environmental aspects. Then, methodologies or approaches to compare different environmental aspects are required.

In Life Cycle Impact Analysis, several methodologies have been developed to integrate impacts in different environmental aspects into the same unit, then to compare different environmental aspects. For example, Eco-Indicator99 [9] and LIME [10] integrate different environmental impacts into a few targets for environmental management, called a safeguard subject. The integration approach is shown in Figure 7.7. This approach is considered a "damage-oriented approach" because it is based on the relationships how inputs and outputs influence on mid-point impact categories, and how mid-point impact affected to the safeguard subject

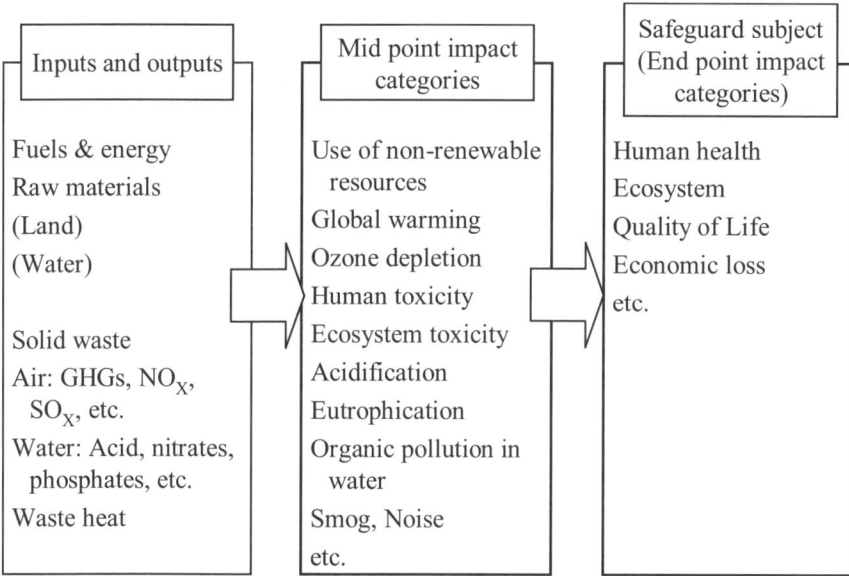

Inputs and outputs	Mid point impact categories	Safeguard subject (End point impact categories)
Fuels & energy	Use of non-renewable resources	Human health
Raw materials		Ecosystem
(Land)	Global warming	Quality of Life
(Water)	Ozone depletion	Economic loss
	Human toxicity	etc.
Solid waste	Ecosystem toxicity	
Air: GHGs, NO_X, SO_X, etc.	Acidification	
	Eutrophication	
Water: Acid, nitrates, phosphates, etc.	Organic pollution in water	
Waste heat	Smog, Noise	
	etc.	

Fig.7.7. Damage-oriented approach for Life Cycle Impact Assessment

Using this approach, we can compare the different impacts of various types of environmental management on the same safeguard subject. The installation of urban wastewater systems gives us one good example. Installation would mitigate some environmental impacts on received water bodies, such as reduction of the risk of infection with waterborne pathogens for those who get in contact with contaminated water, but the construction and operation of these systems have been proven to cause many other kinds of environmental burdens, such as an increase in greenhouse gasses (GHGs), and in hazardous pollutants discharged to air, water and soil. Thus, there is a trade-off between the reduction of health risk by pathogens and the increase of health risk by induced environmental burdens.

This trade-off can be compared by the Microbial Risk Assessment (RA) methodology and Life Cycle Impact Assessment through a quantification of health risk by the Disability Adjusted Life Years (DALYs) unit. Figure 7.8 shows the methodological framework [11]. The RA model was used to estimate the reduction in disease burden while an LCA approach was applied to estimate the environmental burdens affecting human health. The concept of DALYs was used to quantify human health risk in both methods. DALYs is one of the indicators for measuring aggregate health losses, and combines years of life lost with years lived with a disability and is standardized by means of severity weights [12].

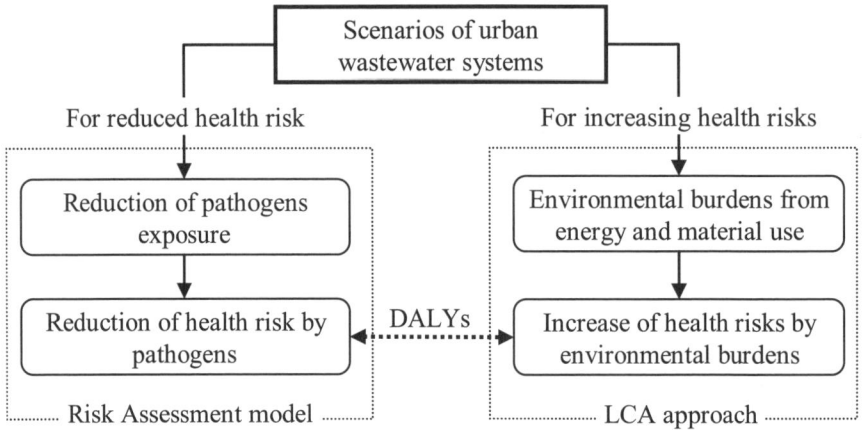

Fig.7.8. Methodological framework to compare reduced and increasing health risks [11]

The following example shows the application result for a hypothetical watershed in which two communities are located along a river [11]. The upstream community discharged their wastewater to the river, and the river water was used as the water supply for the downstream community. Each community was assumed to consist of 200,000 people. The downstream community was assumed to have a conventional water treatment system using rapid sand filtration.

Table 7.2 shows the estimated reduction of health risk by three representative water-related pathogens. The absence of wastewater treatment in the upstream community would cause 3.0×10^{-4} DALYs per year, resulting in a total disease burden for the downstream community of 60 DALYs per year. On the other hand, if we assume the presence of wastewater treatment in the upstream community, disease burden would become 2.8×10^{-5} DALYs per year, and the total disease burden for the downstream community would be 5.7 DALYs per year. Thus, reduced DALYs in the downstream community would total 54DALYs per year.

Table 7.2. Estimated disease burden per person (DALYs/ year)

	Cryptosporidium	Campylobacter	Rotavirus	Total
Without treatment	1.5×10^{-5}	4.3×10^{-5}	2.4×10^{-4}	3.0×10^{-4}
With treatment	1.5×10^{-5}	9.0×10^{-6}	4.1×10^{-6}	2.8×10^{-5}

In order to examine induced environmental burdens, cement and steel consumption for construction of sewer pipes and a treatment plant, the transportation of these construction materials and electricity consumption needed to operate a treatment plant were examined as the major sources of environmental burdens in the life cycle of a wastewater treatment system. The inventory result was converted in terms of the impact to human health by damage factors measured in DALYs/kg-emission, as adopted from the Eco-indicator 99 methodology database [9].

Table 7.3 summarizes total damage from each process considered in both the construction and the operation phases of a sewage management system. For electricity consumption, we could expect a large variation due to the fact of many different energy sources (coal and natural-gas combined cycle) for grid electricity production. Comparing this table with Table 7.2, the reduced health risk by pathogens was higher than the increasing health risks by induced environmental burdens. However, the difference was not very significant, so this result might change depending on the assumption of parameters used for this analysis.

Table 7.3. Health risks by construction and operation of wastewater system (DALYs per year)

Construction phase	
Sewer pipes	1.01
Treatment plant	0.39
Transportation	1.3×10^{-3}
Operation phase	
Electricity	0.50-20
Total	1.9-22

7.5 Holistic assessment of various environmental aspects

We can compare different environmental aspects affecting the same target for environmental management, as in the above example. In many cases, however, we need to compare different environmental aspects relating to different targets of environmental management. For example, if water quality improvement in a nearby lake requires energy and resource consumption, then we need to compare improvement of quality of life (QoL) for nearby residents with the increasing global or regional human health impact and economic loss by induced environmental burdens.

There are basically two approaches to compare different targets of environmental management. One is to convert all aspects or targets of environmental management into a monetary value, then compare them. The other is direct weighting of different aspects or targets of environmental management, either by expert opinions or citizen preferences.

For the former approach, there are a variety of methods of estimating the economic value of the environment [13]. Basically these methods are categorized into revealed preference methods and stated preference methods. Revealed preference methods make use of the actual behavior of people, and include travel cost demand models, hedonic property value models. Stated preference methods draw their data from people's responses to hypothetical questions, so it can be widely applied for valuation of all environmental aspects. The Contingent Valuation Method (CVM) is one of the most frequently used methods among stated preference methods. An individual's willingness to pay (WTP) for the improvement of environmental aspects can be elicited through questionnaire or interview surveys.

Figure 7.9 introduces you to one example of applying these economic valuation methods for the holistic assessment of environmental performance, in this case, lake water quality management [14]. Process steps on the left side of the figure show the approach of economic valuation of lake water quality improvement. After estimating the degree of water quality improvement by each measure using simulation models, the CVM for nearby residents was applied to obtain a monetary value of water quality improvement. Because the value of water quality is different in each lake or water body depending on how citizens use it or its importance to citizens, CVM is a suitable method for economic valuation considering target characteristics through a questionnaire survey of local citizens. Process steps on the right side of the figure show the approach of economic valuation of global warming caused by induced CO_2 emission according to measures of water quality improvement. Life Cycle Inventory Analysis and the marginal cost of impacts by global warming caused by unit CO_2 emission were used to obtain its monetary value. This example was applied into mitigation scenarios of water quality improvement in Lake Suwa, Japan, such as by expanding and improving the treatment of the sewage system, sediment trenching, and controlling non-point source pollution. The monetary value of water quality improvement was one order higher than the monetary value lost by induced CO_2 emission in this analysis.

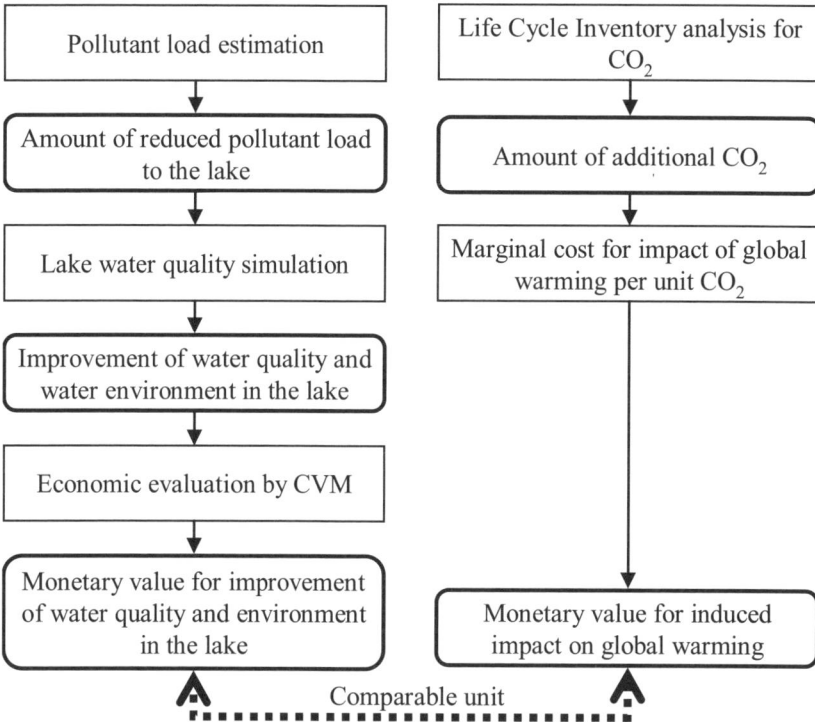

Fig.7.9. Example of application of economic valuation for holistic assessment of lake water quality management

For the latter approach of direct weighting of different environmental aspects or targets, there are several ways to give the weights, such as the Analytical Hierarchical Process (AHP) and choice experiments. AHP is a statistical approach to give the weights through questionnaire and interview surveys [15]. Choice experiments, often called choice-based conjoint analyses, can be categorized into stated preference methods in economic valuation, and can potentially outperform CVM with regard to estimating an individual's marginal willingness to pay (MWTP) for each environmental impact. This can then be converted to the relative importance of the different types of environmental aspects. In brief, a choice experiment asks individuals to choose the most preferred alternative in each choice set. Each alternative consists of several attributes and several choice sets are presented to each individual [16]. If one of the attributes has a monetary price, then it is possible to judge the respondents' MWTPs for the other attributes on the basis of the responses.

Figure 7.10 shows the example of choice experiment applied in the holistic assessment of source separation, collection and treatment of municipal solid waste [17]. Five aspects were chosen, such as health impacts due

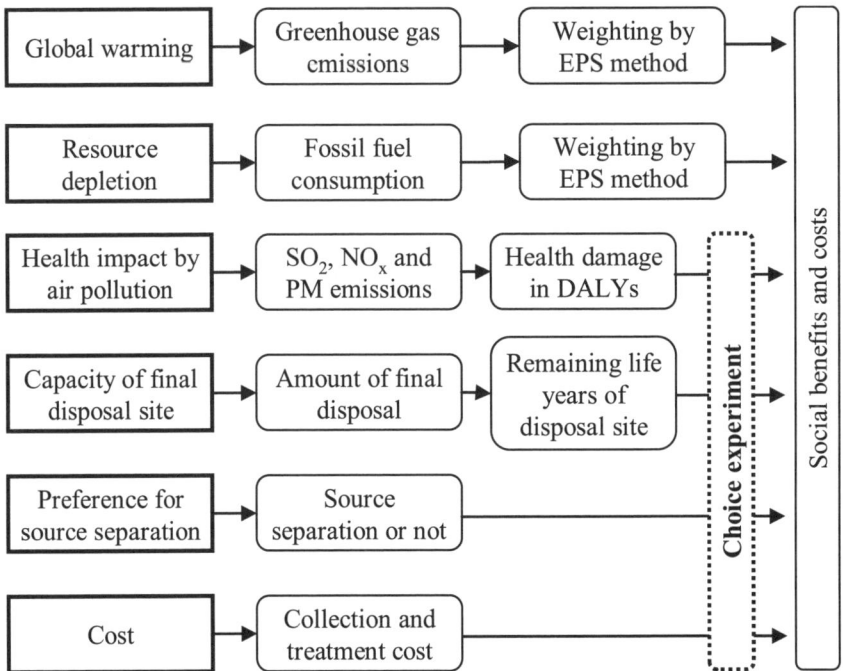

Fig.7.10. Application of choice experiment for holistic assessment on municipal solid waste management.

to local air pollution, global warming, resource depletion, the capacity of the final disposal site, and people's preference for source separation. Among these, the health impacts of local air pollution, the final disposal capacity and people's preference for source separation were analyzed with costs by choice experiment. Next they were compared with each other, and finally their monetary values were estimated to compare global warming and resource depletion. This framework was applied to Kawasaki, Japan, and the choice experiment was carried out through questionnaires to the citizens. Table 7.4 shows the estimated social benefits for these aspects. These values can be used for calculating social benefits for each alternative of municipal solid waste management, allowing for a more holistic consideration.

Table 7.4 Social benefits for each environmental aspect.

	Lower 95%	MPV	Upper 95%
Global warming [million-yen/t-CO_2] [a]	0.04	0.015	0.118
Resource depletion [million-yen/t] [b]			
Oil	0.035	0.071	0.138
Coal	0.002	0.007	0.024
Natural gas	0.034	0.154	0.484
Health impact [million-yen/year] [c]	2,134	3,425	4,615
Final disposal site [million-yen/year] [d]	2,710	4,695	6,552
Source separation [million-yen/year] [e]	3,608	5,690	7,995

MPV: Most probable value
a: Social benefit by 1t-CO_2 decrease of greenhouse gas emissions
b: Social benefit by 1t decrease of each type of fuel consumption.
c: Social benefit by halving health impact (in a unit of DALYs) of air pollutants for citizens in Kawasaki in a year.
d: Social benefit by e times increase of the expected life years of final disposal site in Kawasaki. (e is the mathematical constant.)
e: Social benefit for source separation of plastic waste

7.6 Summary

This chapter introduced several concepts, tools and approaches for holistic assessment of environmental impact by environmental management measures. Most of them are still in the developing phase in terms of methodology and still need to be improved through practical applications. Uncertainties included in the assessment results are also an important issue that we should keep in mind. However, these concepts or approaches will be important for the sustainable development of our society. They will be further studied and widely used in actual situations of planning and assessing alternatives in urban environmental management.

References

[1] Graedel TE, Allenby BR (2003) Industrial ecology, Prentice Hall, New Jersey

[2]   Ministry of Environment, Japan (2003) White paper on Sound material-cycle society (in Japanese), FY2003 edition

[3]   Brunner PH, Rechberger H (2003) Practical handbook of material flow analysis, Lewis Publishers, New York

[4]   Brunner PH, Baccini P (1992) Regional material management and environmental protection, Waste Management and Research, 10:203–212

[5]   Binder C, Schertenleib R, Diaz JH, Bader HP, Baccini P (1997) Regional water balance as a tool for water management in developing countries, Water Resources Development, 13(1):5–20

[6]   Aramaki T, Thuy NTT (2006) Estimation of nitrogen and phosphorus flows by Material Flow Analysis in Haiphong city, Vietnam, Proc. 34th Annual Meeting of Environmental Systems Research, Japan Society of Civil Engineers, pp 45-52

[7]   Wenzel, H. (1997). Environmental assessment of products 1997-1998. Michael Hauschild and Leo Alting. London. Chapman & Hall

[8]   Mendes MR, Aramaki T, Hanaki K (2004) Comparison of the environmental impact of incineration and landfilling in Sao Paulo City as determined by LCA, Resources, Conservation and Recycling, 41(1):47-63

[9]   Goedkoop M and Spriensma R (2001) The Eco-indicator 99 – A damage oriented method for Life Cycle Assessment- methodology report. PRe Consultants B.V.

[10] Istubo N, Inaba A (2004) LIME -A Comprehensive Japanese LCIA Methodology based on Endpoint Modeling-, Proc. of the 6th International Conference on EcoBalance, pp 87-88

[11] Aramaki T, Galal M, Hanaki K (2006) Estimation of reduced and increasing health risks by installation of urban wastewater systems, Water Science and Technology, 53(9):247-252

[12] Murray CJL, Lopez AD (1996) The global burden of disease; a comprehensive assessment of mortality and disability from disease, injury, and risk factors in 1990 and projected to 2020, Global burden of disease and injury series, vol I, WHO/Harvard University Press

[13] Freeman III AM (2003) The measurement of environmental and resource values: Theory and Methods, Resources for the Future, Washington DC

[14] Inaba R, Nakatani J, Aramaki T, Hanaki K (2002) Integrated assessment of environmental improvement in Lake Suwa by multiple countermeasures and impact of global warming by additional CO_2 Emission (in Japanese), J Japan Society on Water Environment, 25:635-640

[15] Saaty TL (1980) The Analytic Hierarchy Process: Planning, Priority Setting and Resource Allocation, McGraw-Hill

[16] Carlsson F, Martinsson P (2001) Do hypothetical and actual marginal willingness to pay differ in choice experiments? Application to the valuation of the environment, J Environmental Economics and Management, 41:179–192

[17] Nakatani J, Aramaki T, Hanaki K (2007) Integrated assessment based on cost-benefit analysis considering the preferences for multi-aspect impacts (in Japanese), Environmental Science, 20: in press

8. Risk Assessments in Urban Environment

Kensuke Fukushi

Integrated Research System for Sustainability Science, The University of Tokyo
7-3-1, Hongo, Bunkyo-ku, Tokyo, 113-8654, Japan

8.1 Introduction

Environmental engineering uses the concept of risk assessment to determine risks of adverse effects on people caused by environmental contamination. The perception of risk is emphasized in this chapter and like conventional risk assessment, a basic concept used to decide environmental regulations or other legislation, there are many conflicts and inconsistencies that raise people's perception of risk. In terms of decision making, the perceived risk is sometimes more important than scientifically determined risk. In order to study the perception of risk, social science, social psychology or other academic fields may be required.

There are many risks concerning urban air quality. Unlike water, foods or many other things, people cannot choose which air to breathe. That is why urban air quality has to be managed properly by the government. Risk caused by pollutants from vehicle exhaust is not that significant in terms of lethal toxicity. Although vehicle exhaust is the most dominant polluting source in the urban air environment, its carcinogenic risk is very low compared to that of smoking and its overall risk is less than that of traffic accidents.

Considering the relatively low risk nature of automobile exhaust, there might be hesitation in making efforts to solve urban air quality problems. However, good air quality is one of the most important livability aspects for people in an urban area. This is proved by the fact that poor air quality is one of the major complaints about the environment in Japan, along with odor and vibration/noise problems. These three complaints make up about 70% of the total environmental complaints in Japan. Odor is considered as human perception of air quality and such parameter may influence public opinion.

Perceptional and scientific determination of risk has to be investigated together. Unfortunately, however, these two concepts are often considered separately. Some examples of differences between actual risk and perceived risk are given below.

Table 8.1 illustrates the perceived risk ranking of certain actions. Researchers asked four groups of people to rank the risk of 30 activities and technologies. The four groups consisted of League of Women Voters (LOWV), college students regarded as the general public with a somewhat higher level of education, active club members regarded as environmentally conscious people, and risk assessment experts. The LOWV and college students ranked the risk of nuclear power the highest. People are obviously afraid of nuclear power. In reality, however, there have been no big accidents in Japan in the last five years. The death toll is only two. On the other hand, many people have been killed by motor vehicles. The national police office in Japan estimates that approximately 8,000 people are killed each year in traffic accidents. Experts ranked nuclear power 20th, a very low rank, while they ranked motor vehicles the highest. In Japan, the highest cause of death is cancer. But for accidental deaths, traffic accidents are quite familiar for the most of people in urban society. As a whole, there is a wide discrepancy between how ordinary people and experts assess risk.

Table 8.1. Ordering of perceived risk for 30 activities and technologies[a] (reproduced from Slovic, 2000)

	Group 1: LOWV	Group 2: College Students	Group 3: Active Club Members	Group 4: Experts
Nuclear power	1	1	8	20
Motor vehicles	2	5	3	1
Handguns	3	2	1	4
Smoking	4	3	4	2
Motorcycles	5	6	2	6
Alcoholic beverages	6	7	5	3
General (private) aviation	7	15	11	12
Police work	8	8	7	17
Pesticides	9	4	15	8
Surgery	10	11	9	5
Fire fighting	11	10	6	18
Large construction	12	14	13	13
Hunting	13	18	10	23
Spray cans	14	13	23	26
Mountain climbing	15	22	12	29
Bicycles	16	24	14	15
Commercial aviation	17	16	18	16
Electric power	18	19	19	9
Swimming	19	30	17	10
Contraceptives	20	9	22	11
Skiing	21	25	16	30
X-rays	22	17	24	7
High-school and college football	23	26	21	27
Railroads	24	23	20	19
Food preservatives	25	12	28	14
Food coloring	26	20	30	21
Power mowers	27	28	25	28
Prescription antibiotics	28	21	26	24
Home appliances	29	27	27	22
Vaccinations	30	29	29	25

[a]The ordering is based on the geometric mean risk ratings within each group.
Rank 1 represents the most risky activity or technology.

Risk can be defined in two ways: the probability that an outcome will occur, or probability that an outcome will occur multiplied by the consequence. Consequence means the severity of the outcome. For example, for the same end result of an infection, the way the infection occurs matters a great deal. If infected by enteric viruses, one might get diarrhea for a couple of days, but if infected by cholera, one can suffer very serious dehydration. Those consequences are very different in their severity. Therefore, risk is usually defined with reference to the consequences involved, though defining the consequences can also be very difficult.

8.2. Determination of risk of pathogen infection

The determination of the dosage-response relationship is illustrated in Fig.8.1. The horizontal axis is the dosage and the vertical axis is the probability of infection. The relationship is easy to understand. The more exposure you have to a risk source, the more likely you are to get infected. In some cases the relationship will be different depending on the kind of pathogens. In order to determine risk, the dose of the pathogen must be determined. If you want to determine the risk of drinking water, you have to know how much tap water you drink a day as well as the amount of pathogens in the tap water. Risk assessment engineers usually assume that people drink on average two liters of water per day, independent of age or gender. Most of the regulations regarding risk are calculated based on this assumption. Perhaps two liters may sound unrealistic for most people in Japan, however, it may be an accurate amount for people in tropical countries.

The determination of pathogen dose in water can be extremely difficult. For instance, imagine a situation where you are asked to determine the cholera infection risk of drinking water in a village suffering a cholera outbreak. During the outbreak, you cannot visit the site. You can go only after it is settled. Because of the time gap, it is difficult to know the actual dose of the water which the infected people took. Even worse, the heavily infected may already be dead by that time, and it is impossible to know their actual dose.

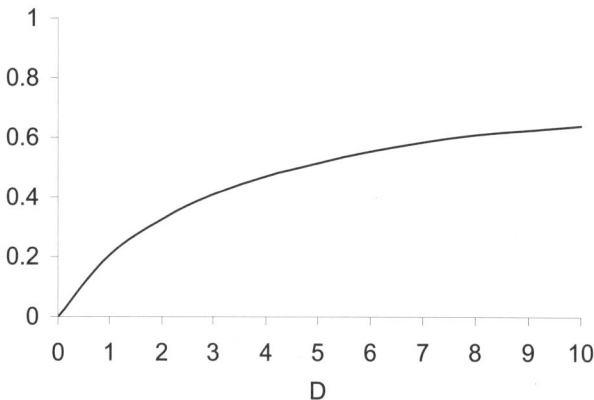

Fig. 8.1. Conceptual dose-response relationship
The graph was drawn based on Poliovirus 3 on Single-hit model.

There are many factors which influence infection risk, such as age, nutrition, and even religion. How does religion correlate with infection risk? Religion itself does not have anything to do with infection risk directly, but it surely determines some habits of a person's daily life. Among all factors, sensitivity to illness is a very important factor. Fig. 8.2 shows relative sensitivity to enteric infection in Japan. The dark-color bars show age groups with above average sensitivity, and the pale-color ones represent age groups that are relatively resistant to infection. Compared to other age groups, the age group of 0 to 4 has an almost three times higher infectious sensitivity risk. Furthermore, sensitivity is also relevant to the severity or intensity of the illness if there is an infection.

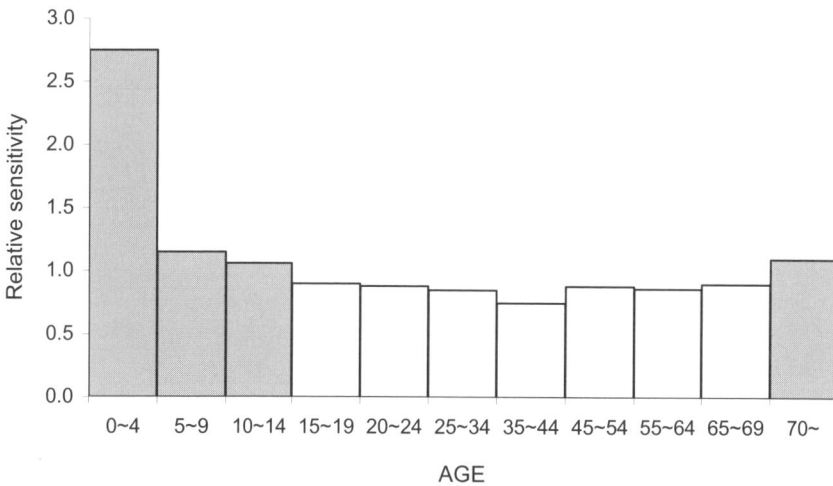

Fig. 8.2. Relative sensitivity of Japanese for enteric infection (reproduced from Watanabe et al., 1999)

In some cases, people may not feel ill, even though they are infected. Such infection is called "unapparent infection." A simulation was carried out to estimate the risk of infection, illness and death by some pathogens for people drinking river water in Bangkok (unpublished data by the author's research group). Actually, nobody drinks river water directly, but this simulation shows how many people out of 10,000 would get infected, ill or die with pathogens such as Hepatitis A, Poliovirus and Rota virus if they drink river water in Bangkok. The simulation results shown in Table 8.2 were obtained by running a program as indicated in Fig.8.3 that applied measurement data of the three viruses in the river water. This program is a very simple one, and adopts the double Monte Carlo method to produce a weighted random number. As for Hepatitis A, around 5,000 people would get infected in a year in this scenario, though this virus often results in un-apparent infection, that is, infection without any obvious symptoms. This characteristic is even more pronounced with the Polio virus. Polio carries a high infection risk in this scenario, with around 7,500 infected in a year, however only about one percent of them would become ill. A much small-er number of people would die from the infection, which is unlike the case of Hepatitis A, in which some people would really die if everybody drank the river water. Despite the low rate of overall illness, if infants become in-fected with the Polio virus, the symptoms will be very serious.

Table 8.2. Simulated numbers of patients infected by various kinds of viruses through river water in Bangkok (10,000 people for one year)

Pathogen	Infection	Illness	Death
Hepatitis A	4706	3530	21
Polio virus	7588	76	0 (0.008)
Rota virus	7773	3885	5

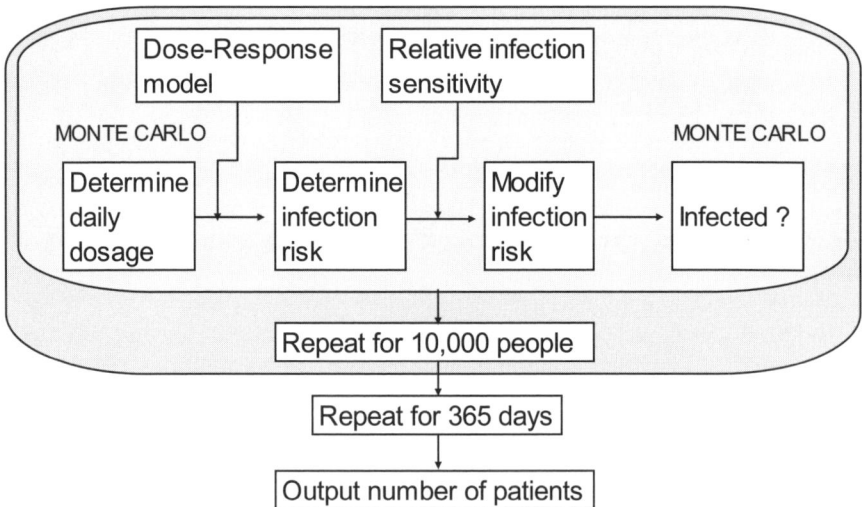

Fig. 8.3. Simulation algorithm for infectious risk calculation

From the numbers obtained from the simulation, the required disinfec-tion efficiency, or virus removal rate, of drinking water can be determined. In reality, however, most of Bangkok's people do not drink river water, even more; tap water, even though water officials declare that the water of Bangkok is safe.

8.3. Case of U.S. beef problem in Japan

This is another example of risk calculation concerning the conflict over the U.S. beef imports from the U.S. to Japan. Nakanishi (2004) calculated how serious or dangerous American beef is, and concluded that overall it is quite safe. Creutzfeldt-Jakob disease (CJD) is caused by abnormal prions from Bovine spongiform encephalopathy (BSE) infected cattle, which are a type of protein and can convert healthy brain cells into nonfunctional ones. As there is no method of curing this disease, it is a big fear for the general public. If people are infected, they first suffer brain-related illnesses, and then die. For American beef by 0.1% examination rate, the risk of infection of CJD was calculated at 1/12 billion by Nakanishi (2004) . Is this low or high? Judging from the U.S.'s Environmental Protection Agency's acceptance level of lifetime risk being 10^{-7}, and 10^{-4} for annual risk, the risk of 1 in 12 billion is extremely low. In case of Japanese beef by 100% examination, the risk is 1/60 trillion. This means that in the case of Japanese beef, we expect one case of CJD every 60,000 years. Even in the case of Japanese beef without any examination and without an attempt to eliminate the most risky portions of beef such as brains, the risk is 1 in 36 billion, which is also very low. What about the death risk of tap water in Japan? It is less than 1 in 3 billion (based on the fact that there has not been a single case of death obviously caused by tap water for the last 30 years in a population of 100 million), but this is still ten times more dangerous than eating Japanese beef.

Returning to the CJD risk, there is a natural occurrence of CJD disease at a 1 in 1 million probability, and this is independent of region, age, race, etc. Japan experiences around a hundred cases of CJD every year regardless of the problem of BSE infected cattle. The facts mentioned above support the position of U.S. officials and beef exporters. Japanese side, by knowing this conclusion from the risk calculation, still insists on complete (100%) examination of beef to confirm "zero risk." A hundred percent examination would decrease the risk by around ten times. This is not a big difference but there is still an emotional push for reducing this risk where possible. Risk perception is quite different from the scientifically, or logically, determined risk. However, voice from public is very important for the government although they are illogical.

8.4. Various expressions of risk

One problem of the conventional definition of risk is that we cannot compare different kinds of risk, such as death risk and infectious risk in the cases of traffic accidents and Polio virus infections. These events have different endpoints. In other words, we cannot say a Polio virus infection is more dangerous than a traffic accident just because it carries a higher risk. The endpoint of a traffic accident is death or injury, while that of a Polio virus infection in most cases is diarrhea, and, in principle, we cannot compare these two endpoints. However, both certainly exert an influence on people's lives, and in daily life we always compare them in some way. Namely, we are always doing risk trade-offs over risk conflicts, or between risk and benefit. For instance, in order to drink water, disinfection of water is necessary, because raw water contains a risk of infection by pathogens. However, disinfection itself produces carcinogenic by-products. Thus, we face certain health risks whether we disinfect water or not.

Meanwhile, the idea of loss of life expectancy (LLE) makes it possible to compare risks that have different endpoints. LLE is an average amount by which one's life is shortened by a particular risk. For instance, a forty-year Japanese male has a remaining life expectancy of 38.4 years. If he is exposed to 1% fatal risk, or death risk, that might cause loss of life expectancy by 0.384 years. This does not mean his lifetime is shortened by this as an individual, however. Suppose one million 40-year old males are collected and exposed to 1% fatal risk, some people may die ten years earlier, while others may live the same lifetime, yet the average lifetime is going to be shortened by 0.384 years.

Table 8.3 is the risk ranking in Japan in order of the length of LLE. Smoking has many endpoints, such as death, heart attack and cancer, and ranks at the top. The value of LLE by smoking varies from several years to fifteen years. The LLE by exposure to diesel particulate matter is 14 days. The LLE by dioxin is 1.3 days. As for the uptake source of dioxin in Japan, fish accounts for about 70-80% of the total, while incineration plants are a minor source. People sometimes misunderstand these particular sources of risk. The LLE by cadmium exposure is 0.87 days. Cadmium comes from many sources, but the major uptake source is food which contains remnants of fertilizer, rainwater, and many other cadmium sources. The LLE by benzene is 0.16 days. Benzene mainly comes from gasoline exhaust.

Table 8.3. Loss of life expectancy (LLE) in Japan (modified from Nakanishi, 2004)

Incident	LLE
Smoking (all)	Several ~ 10+ years
Diesel particulate	14 days
Dioxins	1.3 days
Cadmium	0.87 days
Benzene	0.16 days

Disability adjusted life years (DALY) is another expression of risk. This is employed by the World Health Organization (WHO) for assessing infection risk. It is a measure of premature deaths and losses due to illnesses and disabilities in a population. DALY is calculated by the years of life lost (YLL) plus years lived with disability (YLD). YLL is actual lifetime shortened by some incident. YLD is relatively sensitive and difficult to determine. Judgment of degree of disability is determined by an expert group of medical doctors. For example, if you have a problem with your arm, like limited movement, a doctor will judge how much your life is hindered by the disability. In DALY, a life year for 1,000 healthy people has been set as equally valuable as one life year for, for example, 2,660 blind people, or 1,499 deaf people. It means those numbers of people have the same value of life. But who should decide the value of life? I believe it should not be the government, but possibly an individual or God. Therefore, the concept of valuing lives is sometimes considered problematic by the public.

Determining a value for quality of life is also problematic. Quality of life determines that one with a disability has a lower quality of life than a healthy person, but there is the similar problem of disagreement over how we judge it. As none of these expressions of risk, i.e., DALY or QOL, is supported by the public, it is difficult for a government or an international organization to use them to evaluate risk. Under the present circumstances, some people use LLE, and others use DALY. However, some scientists consider those sensitive discussions or philosophical things required in DALY as scientifically weak and avoid them. On the other hand, mathematically determined risk is of limited use in these sensitive discussions. Actually, there are some people who try to convert those mathematical risk values based on their own view, though I think this is the wrong approach.

8.5. Risk, cost, and benefit

Fig. 8.4 shows the relationship between risk/probability and cost. When we accept a high level of risk, the relative cost for prevention becomes low. However, if you want to prevent damage, you have to spend a lot of money. If you seek zero risk, it is impossible to know how much money you have to spend. In order to optimize risk and cost, a compromise is made by examining the crossing point of the two curves in the figure. Insurance companies frequently analyze the optimization of premium, insurance and risk in this way. Environmental engineers usually decide regulations by taking the compromise point. But sometimes a lower risk than the optimum level is suggested, because people prefer extremely low risk.

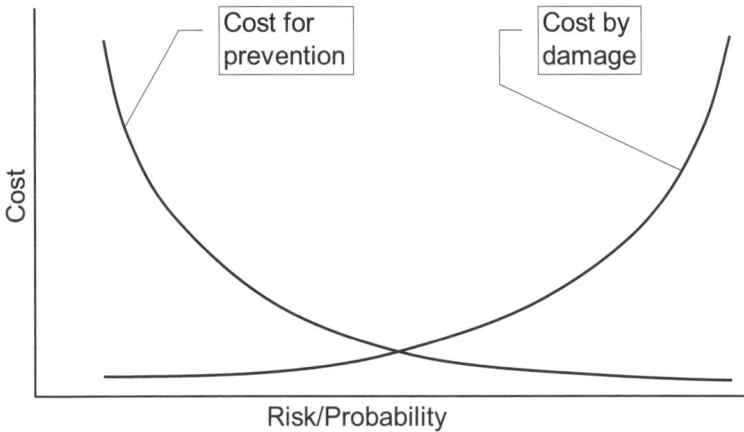

Fig. 8.4. Risk and cost

Table 8.4 shows perceived risk and benefit of various activities, substances, or technologies expressed by scores obtained from questionnaires. For example, people think nuclear weapons are very risky and feel that there is not much benefit received from them. On the other hand, alcoholic beverages, despite their moderately high risk, are considered to have certain benefits, if we control consumption. Motor vehicles are also considered risky and dangerous, but they are very convenient and time-saving for door to door travel, and sometimes essential, especially in places like the U.S. where there are many areas inaccessible by public transport. In case of surgery, cosmetics or solar electric power, people think they receive more benefit than risk. Of course, this is not scientific data. It is based on public perception. But in practice, what politicians and other decision makers listen to is the voice of the public, which comes from the people's perception. Technologies, such as automobiles, surgery, medicines, pesticides and others bring benefits to people. They also bring risk, however. In daily life, people are judging the trade-off between risk and benefit. For instance, the risks and benefits of pesticides and surgery are well understood.

Table 8.4. Perceived risk and benefit (reproduced from Slovic, 2000)

Activity/substance/technology	Perceived risk	Perceived benefit
Nuclear weapons	78	27
Handguns	76	27
Pesticides	71	38
National defense	61	58
Alcoholic beverages	57	49
Motor vehicles	55	76
Asbestos	51	51
Surgery	48	64
Cosmetics	20	49
Solar electric power	12	56

8.6. Risk trade off

The first example of risk trade-off is disinfection of drinking water. By disinfecting drinking water, the risk of infectious disease will be decreased. However, this will increase cancer risk in the form of disinfection by-products. The second example is medical check-up by X-ray. It will decrease integrated medical risk but increase cancer risk. Table 8.5 shows the risk trade-off between infectious risk and cancer risk of drinking water disinfection in accordance with the chlorination intensity. As the chlorination intensity is set higher, the infectious risk is lowered; however, the cancer risk from by-products, such as halogenated hydrocarbons, is elevated. By adding up the values of infectious risk and cancer risk, the total risk is obtained. The optimum chlorination intensity is determined to be around 120 to 600, taking the range of the lowest total risk according to the table. However, the chlorination intensity, or CT value, of 400 to 600 would be too high, and the water would not be drinkable. Therefore, usually the optimum CT value is set lower than this.

Table 8.5. Risk trade off (reproduced from Nakanishi, 2004)

Chlorination intensity	Infectious risk	Cancer risk	Total risk
Raw water	10,000	0.000000	10,000
0	144	0.000000	144
60	14.5	1.3	15.8
120	1.5	1.9	3.4
390	0.00014	3.3	3.3
600	0.000000	4.1	4.1

The decision making process, especially in the civil engineering field, consists of three factors: probability, economics and the consequences of risk. The probability and economics of risk are scientifically/statistically determinable, whereas the consequences of risk involve perceptional aspects and are usually provided by social scientists or sometimes by the voices of the public. By considering those three factors and establishing priorities, decision makers reach decisions.

8.7. Conclusion

We always demand lower risk, and our research investigation methods are intended to lower the risk, for example by cleaning water, purifying solid waste or soil, etc. However, the Japanese life expectancy is now the longest in the world, and people's demands are different depending on the individual or on their ages. Besides, people's perception of risk is very important, in my opinion. Therefore, by knowing how people perceive risks and benefits, researchers can prioritize research topics or find the proper way to express findings so that technologies will be publicly acceptable.

References

[1] Junko Nakanishi (2004) Kankyo Risuku Gaku (in Japanese), Nihon Hyoron Sha, (ISBN: 978-4535584099)
[2] Slovic, P. (2000) Perception of Risk, Earthscan Pubns Ltd.
[3] Watanabe, T., Fukushi, K. and Omura, T. (1999) Risk assessment for waterborne disease by poliovirus in city through drinking water. Preprint of 7th IAWQ Asia-Pacific Regional Conference, 370-375

9. Fate of Water-Related Pathogens in Urban Water Systems

Hiroyuki Katayama

Department of Urban Engineering, The University of Tokyo
7-3-1, Hongo, Bunkyo-ku, Tokyo, 113-8656, Japan

9.1 Introduction

Pathogens transmitted via water are some of the most important agents that raise infant mortality rates, which in turn causes high birth rates and a lack of liberty of women in some communities. The social impact of water-borne diseases is very large.

Infectious disease, including Black Death, has influenced human history. For example, the Black Death outbreak in the 14th century in European countries was estimated to result in a 25% reduction of the population (25 million deaths). The people at that time were unaware of the cause, due to the absence of any means to detect pathogens. We now know that the bacterium named *Yersinia pestis,* sometimes carried by fleas on rats or mice, was responsible for this fatal disease. The Spanish Influenza outbreak in 1918 had a great impact on human society at that time and partly led to the termination of World War I. Disease control is sometimes a military concern.

Anthrax, which was used as a nuisance agent attacking the postal system in the U.S. in 2001, was the first microorganism that was proved to cause disease by Robert Koch and Louis Pasteur almost at the same time independently. In 1884, Koch found that *Vibrio cholerae* was the agent causing a cholera outbreak via water. Cholera is thought to have been an endemic in Indian countries, but was spread to the rest of the world in 19th century, causing many tragedies.

Pathogenic microorganisms are too small to be seen by the naked eye. The invention of the microscope led to the inception of a new branch of knowledge—'microbiology', which has been contributing to the elucidation of cause and remedy of infectious diseases since 19th century.

One of the most important roles of the water supply and sewerage systems is to protect public health from waterborne pathogens.

This chapter explains the relation between the history of microbiology and the development of water-related infrastructure, followed by an explanation of the role of environmental engineering in the protection of public health.

9. 2 Basics of waterborne disease

9.2.1 Life cycle of pathogenic microorganisms

There are many types of pathogens in the world. Why do they exist?

If we try to answer this question on an academic basis, we should think about how they have survived so far in the world. In order to formulate ways to combat pathogens, the strategies adopted by the microbes for survival are of special interest.

They have a life cycle, including a phase of increasing numbers and a phase of transmission to another host. They need to migrate to some "host" after multiplication; without this step their life cycle will remain incomplete and the entire population may become extinct. As a result, pathogens are circulating in human society, causing diseases for people.

Medical doctors put emphasis on the diagnostic aspects of pathogens, including the mechanisms of infection or of interaction between the pathogens and hosts. Environmental engineers, on the other hand, must deal with the route of infection. Environmental engineers aim at ensuring a clean environment where the risk of contamination by pathogens is low.

Different routes of infection, such as direct contact, oral uptake, inhalation, and through contact with animals, exist. For example, HIV is transmitted by sexual intercourse. Meanwhile, *V. cholerae* and the norovirus are spread through water and food. Pathogens of diseases like influenza may enter the body through inhalation. Insects such as the mosquito may be vectors of fatal diseases (Yellow fever, Malaria and others).

9.2.2 Water-related pathogens

Water related disease are categorized into five groups, namely waterborne diseases, water-habitat disease, water-washed diseases (caused by lack of water), water- based diseases and water-related, insect vector diseases.

Waterborne diseases

Waterborne diseases include Cholera, Typhoid, bacillary dysentery, infectious hepatitis, Giardiasis and diarrhoea caused by enteric viruses. These diseases are initiated by ingesting the causative agents in foods or water. The symptoms are dependent on the pathogenic agents, especially in the cases of poliovirus and hepatitis infection. Others are mostly diarrhoea causing agents, but other symptoms (nausea, vomiting, etc.) are also sometimes observed.

Water-habitat diseases

There are some pathogens that do not infect via faecal-oral infection route. These include opportunistic pathogens which are prevalent in the environment and can grow outside of the host bodies, such as *Aeromonas, Legionella, Mycobacterium, Pseudomonas* and *Acanthamoeba*. Some of these pathogens are natural inhabitants of certain water environments.

Water-washed diseases

Water-washed diseases include Scabies, skin sepsis and ulcers, Dysenteries and Ascariasis. These diseases can be avoided easily by washing ones skin and hands with water. Therefore, these diseases may be related to water quantity rather than water quality.

Water-based diseases

Water-based diseases include Schistosomiasis, Dracunuliasis and Filariasis which can penetrate into the human body when a person walks in a puddle with bare feet. These phenomena can happen in an area with a water shortage, for instance when young children in bare feet try to drink water in a dirty place. These diseases are also a water quantity issue.

Water-related insect vector disease

Water-related insect vector diseases include Yellow fever, Dengue fever, Bancroftian filariasis, Malaria and Onchocerciasis. These diseases are prevalent in tropical areas. These pathogenic agents can be controlled by management of the water environment, as was proved by Walter Reed during construction of the Panama Canal.

9.2.3 Pathogens of waterborne diseases

Among the five types of water-related diseases, the latter three are partly social and political problems, since they could be suppressed by simple water supply systems and basic education on public health. On the other hand, waterborne diseases and water-habitat diseases are more difficult problems from the viewpoint of environmental engineering.

The pathogens causing waterborne diseases are limited to those able to survive in an aqueous phase. In terms of the efficiency of infection from emission by one host to contact by a new host, the faecal-oral infection route is the most abundant. This infection route is illustrated in Fig. 9.1.

Pathogens are emitted in an order of 10^{10} particles or more in faeces from a patient per day. Most of them will die off in the environment, diffusing and decreasing their number. But only a small number of pathogens are needed for the next infection, depending on the type of pathogen. Sometimes only one unit of a pathogen can cause infection. Normally, since the conditions outside of a host body are not good for growth, the number of pathogens cannot increase in the environment.

The immune system of the body is working against pathogen infection. It is true that immune systems depend on the individual and their health condition. Minimizing the number of pathogens entering into the body is the most important part of health protection.

Due to increases in human density in urban areas, the pathways of the pathogens have most likely multiplied. Human society has developed various means of protection against infectious disease.

Some cultural and religious behaviours, including a clean and tidy way of life that includes washing hands and bathing have been developed mainly to prevent microbial infection. Heating water or food before drinking and eating is also very effective to inactivate microorganisms.

Infrastructure, such as water supply systems and sewage treatment systems, were developed to prevent infectious diseases.

The infant mortality rate and the access to safe water have a significant correlation. Pathogens are usually excreted from humans or warm-blooded animals, and then come back to the mouths of humans via water, flies, hands or food. To prevent this infectious cycle, attention to water quantity and water quality is essential. Washing dishes, foods, and hands with water will prevent the pathogen from entering into the mouth. Appropriate treatment of faeces will reduce the pathogens going out to the environment so that the level of contamination will be reduced.

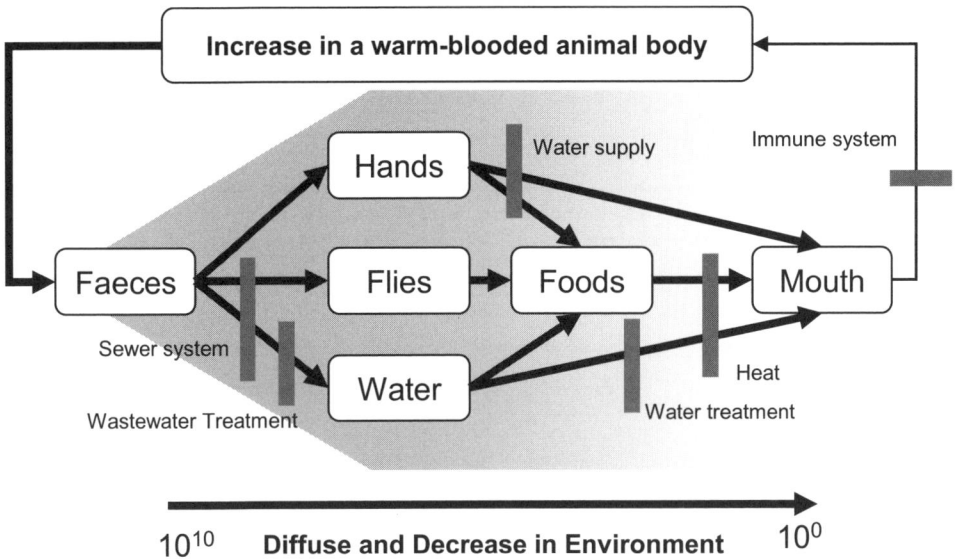

Fig. 9. 1. Life cycle of pathogens of Faecal-oral infection

9. 2. 4 Major agents via faecal–oral infection route

Most diarrheic agents can grow only inside of a warm-blooded animal body. This means that the number of the pathogens decreases in the environment after discharged as faeces. Most of the pathogens will die off in the external environment, but a very few might be able to reach the mouth of their hosts.

There are three groups of the diarrheic agents (Table 9. 1): bacteria, viruses and protozoa. Bacteria are classic waterborne pathogens. They grow in the body of worm-blooded animals and can be inactivated by chlorine. Coliforms or *E. coli* are excellent indicators of faecal pathogens.

The viruses are often associated with waterborne diseases; thus, they are of interest to environmental engineers. Viruses behave differently from *E. coli* since their sizes range 20-100 nm, which is much smaller than *E. coli*.

Protozoa have high tolerance against chlorine, a common disinfectant. *E. coli* is not a good indicator for presence of protozoa for this reason.

The classic bacterial agents have been well managed in developed countries, yet emerging and re-emerging infectious diseases are still causing problems. As for waterborne diseases, emerging and re-emerging agents cause less severe diseases than classic ones in general, except for immuno-compromised people. The socio-economic impacts of infectious disease are still large.

Table 9.1. Features of pathogens to be considered in water management

Pathogens	Resistance to chlorine	Relative infectious dose
(i) Bacteria		
Vibrio Cholerae	Low	High
Campylobacter jejuni	Low	High
Salmonella typhi	Low	High
Shigella spp.	Low	High
(ii) Viruses:		
Adenovirus	Moderate	Low
Norovirus	?	Low
Enterovirus	Moderate	Low
Hepatitis A viruses	Moderate	Low
Rotavirus	?	Moderate
(iii) Protozoa		
Cryptosporidium spp.	High	Low
Giardia intestinalis	High	Low

9. 3 History of Microbiology and Epidemiology

9.3.1 Water as an infectious route

John Snow studied Cholera outbreaks in London in the 19th century. He found that a shallow well was responsible for the high mortality of a street in London in 1849, and stopped the use of the pumping station.

He also compared the prevalence of cholera among different areas, and found that water service areas using downstream of the Thames River as the water source had cholera mortality rates eight times higher than the areas upstream. He concluded that the river water and its consumption as drinking water were responsible for the cholera outbreak. This was the first time that water was identified as an infectious route.

9.3.2 Effect of water treatment

Altona and Hamburg, two neighbouring cities in Germany that both used the Elbe River as a water source, were found to have very different mortality rates during an outbreak of Cholera in 1892. This was explained that Altona provided drinking water after slow sand filtration while Hamburg employed only sedimentation. In response, a sand filtration system was installed in Hamburg to reduce the risk of infection.

In the middle of the 19th century, chlorination began to be used in hospitals for disinfection, and was then added to many public water supplies. The effectiveness of chlorination was widely known in the 20th century and is now employed worldwide.

Coliforms are microbes which co-exist with the faecal pathogens in the intestines of humans as well as other worm-blooded animals. Hence, the absence of coliforms in water indicates the absence of faecal pathogens.

Pipeline systems were also developed and used in water supply systems. Pipelines prevent pathogenic contamination during distribution due to positive pressure applied in the pipe to avoid intrusion of water from outside.

9.3.3 Modern water supply system

In the 19th century, effective water treatment schemes to combat pathogens were established. The modern water supply system consists of the following factors: Coagulation using coagulant (Al, etc) to form floccules, sedimentation to settle the particle matters to the sediment, sand filtration, (rapid sand filtration is widely used in Japan) and chlorination. To ensure the safety, the absence of total coliforms, thermotolerant coliforms or *E. coli* is confirmed.

The wastewater treatment has not been considered in relation to infectious disease, but rather in terms of environmental issues. For example, Japan started to build modern a water supply in the late 1880s, and covered 80% of the population in 1970. Coverage of wastewater treatment systems, which was only 16% of the population in 1970, increased to 69% in 2005. However, wastewater treatment plants also play an important role in preventing the waterborne diseases. The functions of water infrastructures (water supply, sewer system and wastewater treatment) are demonstrated in Fig 9. 1.

9. 4 Removal of Pathogens in Water

9.4.1 Removal efficiency in water purification methods

Role of water purification

The role of water purification is not only to decrease the number of pathogens but also to enhance the effectiveness of disinfection. Removal of suspended solids in this treatment contributes to stable disinfection.

Rapid sand filtration

Rapid sand filtration is used around the world as a standard method of water treatment. During the treatment, approximately 99% of the pathogens are expected to be eliminated if properly operated. In the maintenance stage, reverse washing of the sand is used to renew the clogging effect. Since the washed water contains accumulated pathogens including *Cryptosporidium*, this water should not be recycled as source water.

Membrane filtration

Membrane filtration has become a popular water purification method. According to the pore size, membranes are categorized as MF (micro filtration), UF (ultra filtration), NF (nanofiltration) and RO (reverse osmosis filtration).

During operation, suspended solids in the source water will be rejected by the membrane, together with other impurities including organic matter and metals. Blocking such solids is good from the viewpoint of water quality, but at the same time, this increases the need of maintenance to clear the membranes.

In Japan, the application of chlorine to finished water is mandatory. The most important targets in water treatment are protozoa such as *Cryptosporidium*. Membrane filtration is a perfect barrier against *Cryptosporidium*.

Viruses are also removed by membrane filtration. UF, NF and RO have smaller pore sizes than the viruses and are therefore able to reject them. Viruses are not rejected perfectly by MF, which has a larger pore size than the viruses, yet they are removed with the cake layer that forms on the surface of the membrane (Otaki et al. 1998).

9.4.2 Disinfection

Chlorination

Chlorine is used worldwide due to its effectiveness as a disinfectant, residual effect and availability. The residual effect is important because a part of the pipeline network of water supply sometimes has negative pressure, and thus, contamination by pathogens may occur. The growth of biofilm in the pipeline is suppressed by residual disinfectants.

The pH value of water should not exceed 8.0 for the efficiency of chlorination.

The reaction of chlorine is:

$$Cl_2 + H_2O \rightarrow H^+ + Cl^- + HOCl \tag{1}$$

$$HOCl \leftrightarrow H^+ + OCl^- \tag{2}$$

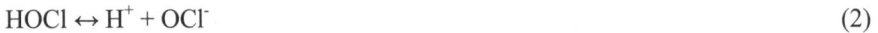

Equation (2) is dependent upon pH. At pH levels higher than 8.0, the dissociated forms (OCl^-) predominate, and at a lower pH, undissociated hypochlorous (HOCl) acid predominates. The dissociated forms have less disinfection efficiency due to the repulsive effect against the surface of microorganisms which are normally slightly negatively charged.

Suspended solids, organic matter and ammonia in water also influence the effect of chlorination.

Disinfection by-products (Trihalomethane, Haloacetic acid etc.) are associated with chlorine's use. Chlorine dioxide and chloramines have been proposed as alternatives to chlorine in this regard. However, chlorine dioxide is suspected to be carcinogenic, and it is not recommended as a disinfectant. Chloramines are less effective in inactivation of microorganisms, and moreover, NDMA, a by-product of chloramines, has been found to be extremely toxic to humans. Therefore, the use of chloramines is also not recommended.

Last but not least, chlorine is not effective against protozoa like *Cryptosporidium.*

Ozonation

Ozonation is effective for the removal of odour and colour of the water, which are the main reasons for the installation of water purification plants in Japan. Ozone is a strong oxidant creating many active species and results in inactivation of microorganisms, though its mechanism has not been fully elucidated. Ozonation is effective against bacteria, protozoa and viruses.

After ozonation, biological activated carbon filtration is usually applied, and some further reduction of pathogens is also expected.

Ultraviolet irradiation

UV light irradiation is one of the most effective disinfection methods against bacteria, viruses, and protozoa in source water and wastewater. UV disinfection systems are easy to maintain, need no additional chemical inputs and produce no hazardous by-products. Therefore, water treatment plants that utilize UV disinfection processes have increased in number in recent decades.

In the 1990s, *Cryptosporidium* was the most important pathogen targeted by the water treatment industry. The viability of *Cryptosporidium* oocysts had been examined by excystation, but in the late 1990s it was examined by infection assay using the sensitive mouse or the specific cell line. The newer methods revealed that UV was very effective in inactivating *Cryptosporidium* (Abbaszadegan et al. 1997).

Inactivation of microorganisms by UV irradiation is effected through the formation of lesions in the genomic DNA of the organisms. However, UV irradiation is not always perfect because some organisms are known to possess the ability to repair their DNA by mechanisms such as photoreactivation and dark repair. *Cryptosporidium* was tested for its abilities of photoreactivation, and it was found that *Cryptosporidium* had no such ability (Oguma et al. 2001).

The Ministry of Health and Welfare, Japan, approved the use of UV irradiation as a barrier against *Cryptosporidium* for water supply systems in 2007.

How to select a disinfectant

The selection of a specific disinfection method often depends on the mode of water use and existing regulations. For instance, the photoreactivation potential of microbes would be concerned in the use of UV treatment where treated water may be exposed to sunlight. Again, in case of wastewater rich in organic matter, a high chlorine dose should not be used in order to avoid the emergence of toxic by-products. Government regulations often make it compulsory to use a certain technology. For example, in Japan, residual chlorine in tap water is prescribed by regulations.

9. 5 Risk management

9.5.1 Indicator of pathogens

Chemical risk agents, such as cyanogens or trihalomethanes, are included in water on the order of ng/ml. This is quite a small value as concentration, but considering the number of particles, these values should be multiplied by Avogadro number (6.02×10^{23}). When related to microorganisms, the number of particles is directly reflected in the observation. In other words, the pathogens are discrete and not in a solution. This is the reason why the number of the pathogens in water varies.

Some chemical agents, such as organic mercury, accumulate in the body. The risk of pathogens, however, is not cumulative. Pathogens that are ingested will either: 1) infect the human, 2) be digested, or 3) simply pass through the stomach. In case of the risk assessment of drinking water, the number of pathogens taken in a day is used to calculate risk of infection.

Since the number of pathogens is highly variable, it is not always reliable to judge the water safety based on the inspection of pathogens. Measuring every kind of pathogen is not practical either.

As for source protection and water quality analysis, the most important aspect to consider is faecal contamination, especially human faecal contamination, of the water. Therefore, among the various water quality parameters, those related to faecal contamination should be emphasized. Characteristics of the ideal faecal indicators are that they:

1. should be present in wastewater and in contaminated water
2. should be present in greater numbers than the pathogens
3. should not multiply in environment
4. should be more tolerant in the environment than pathogens
5. should present no health risk
6. should be easy to enumerate

Among the various parameters, the total coliforms, thermotolerant coliforms and *E. coli* are used as indicators of faecal contamination worldwide because they are present in a great amount in faeces of warm-blooded animals including humans, are easily detectable, do not multiply in the normal environment and have a fate similar to most of pathogens.

As the principal indicator bacteria, *Escherichia coli*, thermotolerant coliforms (95% are *E. coli*) and faecal streptococci are recommended by the World Health Organization (WHO, the webpage of its document is listed in the reference). Bacteriophages are considered to be promising viral indicators.

9.5.2 Risk assessment

Risk assessment for a chemical agent consists of four steps: 1) Hazard identification, 2) Dose response assessment, 3) Exposure assessment and 4) Risk characterization.

Risk is calculated on the basis of the probability of the infection. Of course, the symptoms of the disease are very important. However, in microbial risk analysis, infection is used as the target of the final hazard because the symptoms are variable and, moreover, asymptomatic infection can cause a secondary infection. The risk of secondary infection is difficult to evaluate.

The microbial risk of infection can be calculated by exposure assessment and dose-response assessment. The easiest method of exposure assessment is to calculate the product of the concentration of the microorganism and the amount of water consumed by people. However, the concentration of pathogenic microorganisms is highly variable because the number of particles in water is quite small compared to chemical agents of the same hazard level. Therefore, the estimated concentration does not always represent the risk of infection.

Microbial doses are not cumulative, so the risk of infection has no relation to the uptake in previous days. To compare with chemical agent risk, the microbial risk is usually shown as an annual risk of infection. In a normal case, the expected intake of pathogens is not so high. Therefore, the risk of infection is linear in relation to the intake of pathogens, which is proportional to the annual average (arithmetic mean value) of pathogens (Masago et al. 2006).

In the water supply systems, the concentration of pathogens is not so high, and one is not likely to be exposed to more than one particle of a certain pathogen. Therefore, the possibility of infection due to one pathogenic agent is important, but this figure is not easily obtained from the challenge data. The challenge data are obtained by using volunteers as a host to pathogens, while a known amount of the pathogens are ingested to observe the probability of infection. This method has provided much informative data, yet it has been avoided in recent years due to ethical considerations.

The maximum concentration of the pathogens might be the most important factor in the annual risk of infection.

9.5.3 Risk management

The most normal approach is to regulate the maximum concentration of hazardous agents. However, the normal method is not practical in the case of pathogen control, because the water amount needed is too large to be monitored routinely. Instead, a treatment-based control approach may be effective.

Since an outbreak is an acute event, prevention of the accident is the first priority. Historically, outbreaks always have had some apparent causes, including absence of chlorination or an invasion of untreated wastewater. Prevention of the accident is important in developed countries as well. Countermeasure systems in case of an accident are also important for the public health sectors, as well as disclosure of the information that there has been an outbreak.

A multilateral approach is necessary to develop water safety against microbial agents. Fig 9.2 describes how to select the monitoring point of faecal contamination in water. Hazardous analysis and control critical point (HACCP) is a useful approach, which addresses microbiological hazards as a means of prevention rather than finished product inspection. In the case of the water industry, the water safety process should consist of water quality analysis, source protection and minimum treatment requirements. In the water quality analysis stage, indicators (microbial indicators, turbidity, chlorine, etc.) should be monitored at an appropriate frequency. For source water quality analysis, the most important aspect is faecal contamination, especially human faecal contamination. Among water quality parameters, those related to faecal contamination should be emphasized.

Pathogen analysis is expensive, dangerous for the inspectors and sometimes impractical because the necessary techniques can be time consuming. Reliance is therefore placed on relatively simple and more rapid tests for the detection of certain intestinal bacteria which indicate that faecal contamination could be present.

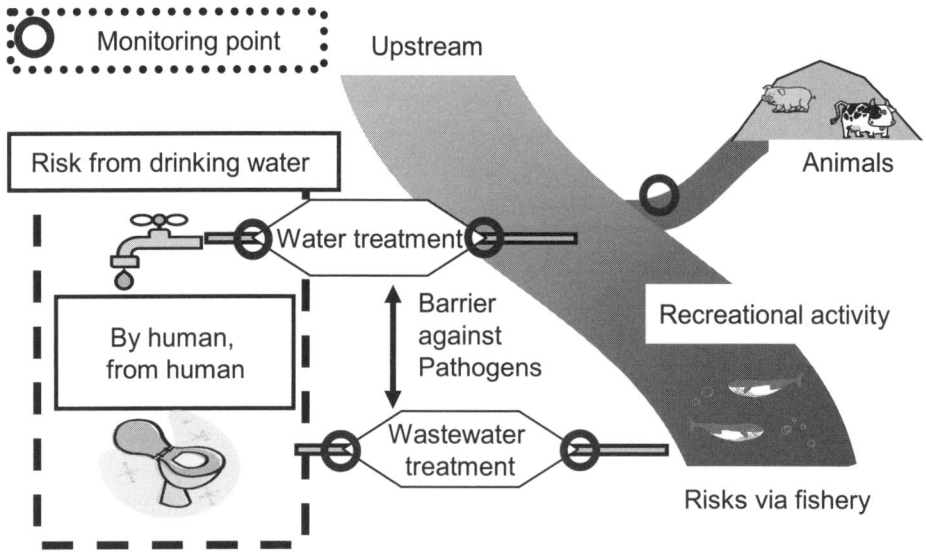

Fig. 9.2. Monitoring point for management of microbial risk

9.6 Topics in recent health-related water microbiology

The classic waterborne diseases were well controlled by the water supply systems and the sewage systems built in developed countries in the 20th century. The basic strategy was to remove particles in the water, followed by disinfection by chlorine, and confirm the efficiency by analysing the presence/absence of coliforms. This worked quite well for the classic bacterial pathogens, but is not successful enough for the protozoa or viruses, which are more tolerant to chlorine.

In the 1970s, the threat of viruses was taken seriously in the U.S. and European countries. However, viral infection via water was found only in small water supply systems of an accidental insufficient process, sporadic cases were not epidemiologically detected in large water supply systems. Consequently, no major countermeasures were employed in most countries.

In 1980s, the disinfection by-products of chlorination were suspected to be carcinogenic. There have been arguments regarding the use of chlorine since then in developed countries, and some have ceased the use of chlorine in the water supply.

In the 1990s, *Cryptosporidium* was of great concern. There was a large outbreak in Milwaukee, USA in 1993, resulting in more than 400,000 cases. In Japan, Ogose town suffered an outbreak of *Cryptosporidium* in 1996 and reported more than 8,000 cases, which is more than 70% of the population in the water supply service area. Cryptosporidium is highly tolerant to chlorine and other chemical disinfectants. *Cryptosporidium* has a small diameter of 5 um, and can pass through sand filtration. Therefore, the risk of infection of *Cryptosporidium* is not reduced sufficiently by conventional water purification processes.

In the 2000s, norovirus was found abundant in the water environment (wastewater, river water, wells, seawater, etc). Some types of seafood (e.g., oysters) often contain this virus. This virus uses human as the sole host. It is highly tolerant to acidic pH, high temperature and chlorine doses (Keswick et al., 1985).

The risk of viral infection via water has become the topic of serious discussions. Although water is not responsible for most of the outbreaks, water is a possible infection routes, and establishing safety procedures and processes is an urgent issue for waterworks. Some European countries, some states in the U.S. and South Africa have started to develop water quality standards regarding viruses in drinking water and recreational water.

References

[1] Abbaszadegan M, Hasan MN, Gerba CP, Roessler PF, Wilson BR, Kuennen R and Dellen EV (1997) The disinfection efficacy of a point-of-use water treatment system against bacterial, viral and protozoan waterborne pathogens. Wat Res 31: 574-582

[2] Keswick BH, Satterwhite TK, Johnson PC, DuPont HL, Secor SL, Bitsura JA, Gary GW and Hoff JC (1985). Inactivation of Norwalk virus in drinking water by chlorine. Appl Environ Microbiol 50: 261–264.

[3] Masago Y, Katayama H, Watanabe T, Haramoto E, Hashimoto A, Omura T, Hirata T and Ohgaki S (2006) Quantitative Risk Assessment of Noroviruses in Drinking Water Based on Qualitative Data in Japan. Environ Sci Technol 40: 7428-7433.

[4] Oguma K, Katayama H, Mitani H, Morita S, Hirata T and Ohgaki S (2001) Determination of pyrimidine dimers in Escherichia coli and *Cryptosporidium parvum* during ultraviolet light inactivation, photoreactivation and dark repair. Appl and Environ Microbiol 67: 4630-4637

[5] Otaki M, Yano K and Ohgaki S (1998) Virus Removal in Membrane Separation Process. Wat Sci and Tech 37(10): 107-116

[6] World Health Organization, Guidelines for drinking-water quality. (Training materials in water, sanitation and health) http://www.who.int/water_sanitation_health/dwq/en/

Part III   Heat and Energy Management

10. Global Climate Change and Cities

Keisuke Hanaki

Department of Urban Engineering, The University of Tokyo
7-3-1, Hongo, Bunkyo-ku, Tokyo, 113-8656, Japan

10.1. Introduction

The development of the human life style and its impact on the environment has changed from the primitive era to the current, modern world. In the primitive era, nature was the enemy to overcome. Human beings were mostly exposed to stress by nature including extreme cold, disasters, floods, diseases and the threat of other animals (Fig. 10.1-a). Mostly, the history of human beings has been one of battling against nature. After the development of human society, humans started to form groups to live together and establish their territories. Humans started to maintain safe environments and obtain food within this territory by cultivating crops and others. Human beings started to influence nature at this stage (Fig. 10.1-b), though overall human impact was still limited. After the industrial revolution, the area needed for humans to develop their activities expanded because more energy was consumed, more food was needed and people started to travel farther. As the land area utilized by humans expanded, so did population. As a result, each city or village began to have a larger zone of influence and the number of cities increased. Nowadays, the area of influence, or human impact, overlaps among cities (Fig. 10.1-c). Nature can be found in only limited places, and now it has to be preserved. Most natural areas are impacted by human activities and increased population. Such human influence is now called "environmental loading," which includes a wide variety of effects by human beings on nature. The emission of pollutants, the use of resources, threats to the ecosystem and land use changes are types of "environmental loading."

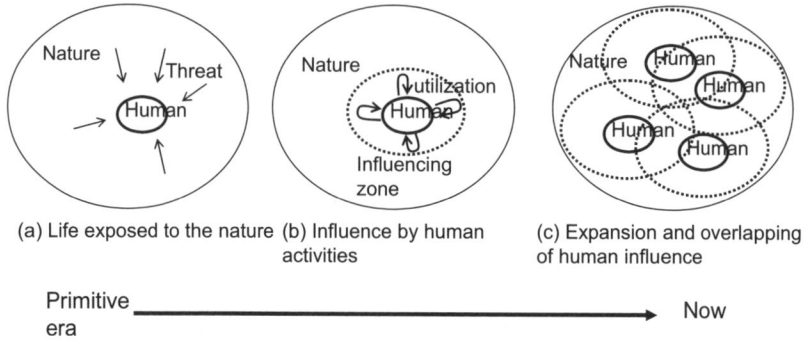

(a) Life exposed to the nature (b) Influence by human
activities

(c) Expansion and overlapping
of human influence

Primitive
era

Now

Fig. 10. 1. Relationship between human beings and nature.

Unlike communities in the old era, contemporary cities do not necessarily contain agricultural or industrial activity inside of them. Instead, cities depend on agriculture and industry outside the urban area. In order to become environmentally sound, cities should reduce their degrees of environmental loading, including energy and material consumption.

However, the reduction of environmental loading alone is not the objective of urban management. Maintaining or even improving the quality of life (QOL) of people is the objective of societies in both developing and developed countries. Satisfaction of basic human needs (BHNs) such as safety, health and education is the main concern in least developed countries. The goal in many newly industrialized countries is to improve the level of physical comfort among people, including better housing and higher automobile ownership rates. Many of these countries are experiencing significant environmental problems such as air pollution and water pollution. The QOL of people in developed countries is changing from physical comfort to mental satisfaction and to environmental consciousness.

Climate change, which is also called global warming, has become the most challenging problem for human beings to solve. Predicted serious climate change in the future is obviously brought about by human activities, but this issue also restricts the activities of human beings, especially in urban areas. Urban areas are the major sources of greenhouse gases (GHG) such as carbon dioxide due to energy consumption and methane emission from solid waste landfills. Cities in developing countries are likely to be main contributors of GHG emissions in the near future. On the other hand, cities in developed countries are expected to become leaders in reducing levels of fossil fuel consumption.

10.2. Impact of climate change on urban activity

10.2.1. Factors determining vulnerability

Climate change causes not only temperature rise, but also a change in precipitation. The sea level rises as the result of temperature rise. The impact of climate change on a system in a particular region is usually determined by three factors, namely system exposure, sensitivity to climate change and adaptability to climate change. An area's level of system exposure describes the intensity of climate change such as temperature or precipitation in the region. These are inputs of climate change. Sensitivity means how the system is influenced by the climate change. If we take the example of rice production in a particular area, system exposure would be the future temperature and precipitation change that potentially causes the problem. Sensitivity depends on whether that particular area is usually hot or dry. Rice production in dry areas is sensitive to future precipitation decrease. Adaptation is the capacity of farmers to change their cultivation practices or the species of rice. These factors determine vulnerability to climate change. Identifying vulnerable areas is important to prepare a strategy for dealing with future climate change.

Water resources are particularly vulnerable to the threat of climate change. Urban areas depend on large amounts of water for domestic as well as industrial purposes, and disruption of the supply of water could have significant consequences. The amount of available water is approximately calculated by deducting evaporation from precipitation. Temperature rise increases evaporation whereas the change in levels of future precipitation is still uncertain. Sensitivity of water management system depends on the balance between water demand and water resources. Population growth and industrial development increases sensitivity and vulnerability. Storing water in reservoirs can increase adaptability. Megacities in developing countries located in dry regions will suffer the most from water shortages because of inherent water scarcity, climate change and demand increases.

Population growth is very rapid in some developing countries. Urban population growth rate in such countries is usually higher than the national population growth rate. Table 10.1 shows the top 10 large city regions in the world in 2000 and as projected for 2015 [1]. The growth rate of cities in developing countries in Asia is high. Seven of the top ten megacity regions in 2015 will be located in Asia. Such rapid growth can cause problems such as poor infrastructure that cannot catch up with the tremendous rate of population growth.

Table 10.1. Population of top ten megacity regions.

	2000			2015 (projection)	
Rank	Agglomeration and Country	Population (millions)	Rank	Agglomeration and Country	Population (millions)
1	Tokyo, Japan	34.5	1	Tokyo, Japan	35.5
2	Mexico City, Mexico	18.1	2	Mumbai (Bombay), India	21.9
3	New York, USA	17.8	3	Mexico City, Mexico	21.6
4	Sao Paulo, Brazil	17.1	4	Sao Paulo, Brazil	20.5
5	Mumbai (Bombay), India	16.1	5	New York, USA	19.9
6	Shanghai, China	13.2	6	Delhi, India	18.6
7	Kolkata (Calcutta), India	13.1	7	Shanghai, China	17.2
8	Delhi, India	12.4	8	Kolkata (Calcutta), India	17.0
9	Buenos Aires, Argentina	11.8	9	Dhaka, Bangladesh	16.8
10	Los Angeles, USA	11.8	10	Jakarta, Indonesia	16.8

Data from the United Nations [1]

Such rapid population growth in large cities is due to the national growth rate and to the migration of people to urban areas. Migration from rural areas is occurring in developing countries at an extraordinary rate. This type of migration pattern still occurs in developed countries as well, as shown in Table 10.2. This trend also suggests that migration to urban areas will continue after 2030 in developing countries.

Table 10.2. Urban population percentage

	1970	2000	2030 (projection)
World average	36.8	47.2	60.2
Developing region	25.1	40.4	56.4
Japan	71.2	78.8	84.8
China	17.4	35.8	59.5
Thailand	13.3	19.8	33.1
India	19.8	27.7	40.9

United Nations [1],   Note: definition of urban population varies among countries.

Since many megacities are located in the coastal zone, rising sea levels could become a threat depending on topographical conditions. A map produced by the IPCC [2] projects the world's vulnerable coastal zones that will likely have both large population and a high susceptibility to damage by sea level rise by 2050 (Fig. 10.2).   Measures to increase the capacity of adaptation, such as construction of dykes to control storm surges, can reduce the risk, but highly populated, flat areas are difficult to protect.

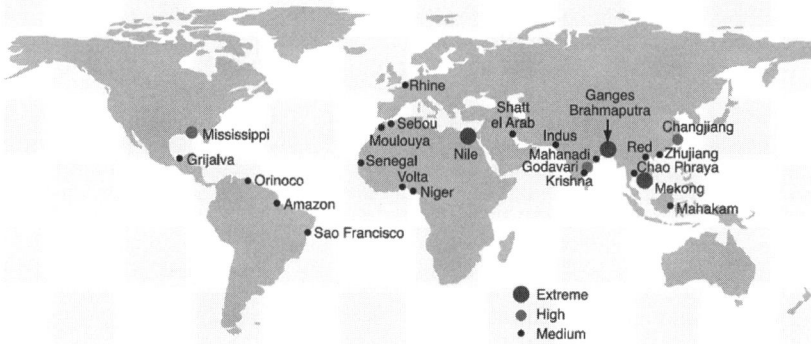

Relative vulnerability of coastal deltas as shown by the indicative population potentially displaced by current sea-level trends to 2050 (Extreme = >1 million; High = 1 million to 50,000; Medium = 50,000 to 5,000; following Ericson *et al.*, 2006).

Fig. 10.2. Relative vulnerability of coastal areas in 2050 [2]

10.3. Emission of greenhouse gases from urban activity

10.3.1. Major sources

There are many varieties of anthropogenic emission of GHG. The most important GHG for global climate change is carbon dioxide, but methane and nitrous oxide are also important gases to control. Urban activities cause GHG emission in both direct and indirect ways. Table 10.3 shows GHG emission caused by urban activities. Fossil fuels support most human activities in industrialized countries. Urbanization and the immigration of people from rural to urban areas often increase energy consumption per person due to the change of lifestyle. People in urban areas use automobiles more, as well as appliances and other machines that require high levels of energy consumption. The second source of carbon dioxide caused by urbanization in developing countries is through deforestation. The demand for more land to live on and to cultivate food decreases the amount for forests in areas surrounding the megacities. This is one of the most difficult deforestation problems to solve. The last type of carbon dioxide emission from urban activities is from the cement industry. Carbon dioxide is released when cement (CaO) is produced from limestone (CaCO$_3$). This type of carbon dioxide emission has increased with greater construction of infrastructure in cities such as roads, sewer pipes, tunnels and buildings. This type of carbon dioxide emission is evaluated by using the Life Cycle Assessment (LCA) method.

Table10.3. Contribution of urban activity on anthropogenic emissions of greenhouse gas.

GHG	Source	Anthropogenic contribution	
		General	Urban area
Carbon dioxide	Fossil fuel	Industrialized society	High consumption with urbanization = energy consumption increase (developing countries)
	Deforestation	Land use change: commercial logging, agriculture and settlement of people	Logging for paper, wood products
	Cement industry		Construction of infrastructure
Methane	Rice cultivation	Increasing food demand	
	Livestock	Increasing food demand	
	Solid waste landfill		Solid waste generation exceeding land's capacity
Nitrous oxide	Fertilizer	Anthropogenic fixation of nitrogen gas	
	Combustion	Biomass burning	Automobiles, solid waste and sewage sludge incineration
	Wastewater		Load exceeding natural capacity
Ozone	Nitrogen oxide & hydrocarbon		Air pollution (automobiles, industry)
CFCs	Industrial products	Industrialized society	

The major source of methane from urban activity is solid waste landfill gas. Biodegradable, organic solid waste is degraded in the landfill site without air and is eventually converted to methane gas. Although most solid waste is incinerated in Japan, most of the developing countries as well as some of the developed countries use landfills for solid waste disposal. Methane recovery at the landfill site and conversion to electricity is often done as a CDM (Clean Development Mechanism) project in developing countries. Other methane sources are agricultural activities that are indirectly enhanced by urban food demand. Nitrous oxide is emitted in agricultural areas by microbial transformation that also takes place in wastewater treatment and in polluted water environments. The other physical-chemical route is nitrous oxide emission with combustion. The catalysts of automobile exhaust gas and incineration of sewage sludge cause its emission in developed countries. Ozone is generated mainly due to automobile air pollution in urban areas. Ozone is also a greenhouse gas, though quantification of its contribution to global warming is difficult due to its short life and heterogeneous spatial distribution.

Carbon dioxide emission is usually categorized into industrial, transportation, commercial/office and household sectors. Total contribution of each sector to carbon dioxide emissions in Japan is: industrial, 45.5%, transportation, 19.9%, commercial/office, 18.4%, residential, 13.4%, and others, 2.8%. It should be noted that the industrial products are used mostly in the urban areas. In other words, material consumption in urban areas induces industrial carbon dioxide emission.

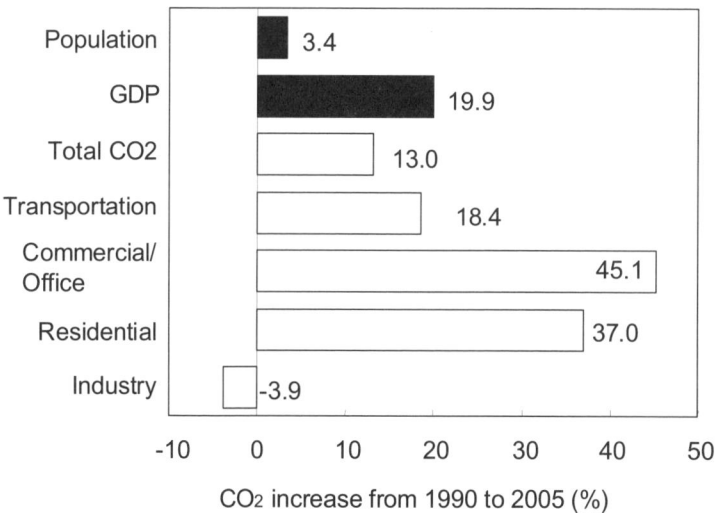

Fig. 10.3. Change in sectors of carbon dioxide emissions in Japan from 1990 to 2005 (%)

Fig. 10.3 shows the recent trends of carbon dioxide emissions for each sector in Japan. Total carbon dioxide emissions in 2005 increased by

13.0% from 1990, the reference year of the Kyoto Protocol. This figure clearly indicates that non-industrial sectors are increasing their carbon dioxide emissions significantly. Carbon dioxide emissions from urban areas are increasing, as well. Japan has declared a goal of reducing annual emission by 6% from the 1990 level in the period from 2008 to 2012. There is no possibility of fulfilling the Kyoto Protocol without controlling non-industrial sectors. Urban activity must be improved so that the carbon dioxide emission is reduced.

10.3.2. Carbon dioxide emission at the urban scale

The most important source of greenhouse gas is carbon dioxide from energy consumption, which depends on various factors. Influencing factors can be divided into two types, namely the circumstances of the country as a whole and the structure of the urban area (Fig. 10.4).

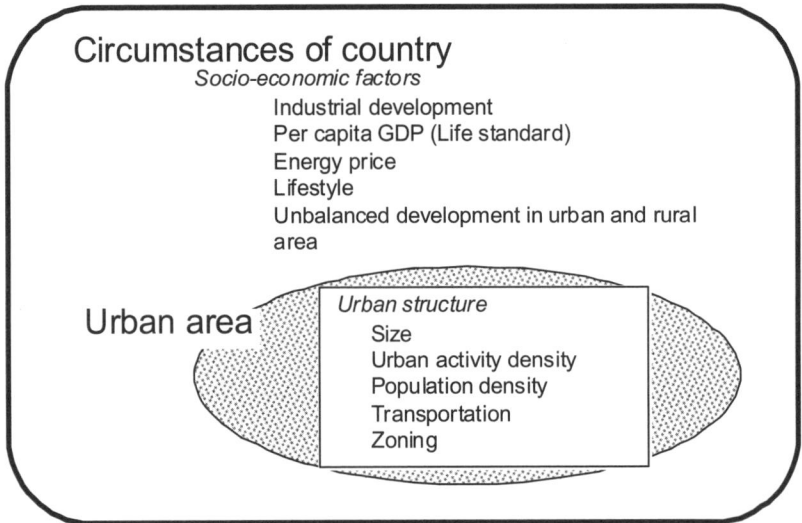

Fig. 10.4. Factors influencing energy consumption in urban areas.

Socio-economic factors in the country are the most significant. Energy consumption levels in developing and developed countries differ because of the different levels of industrialization, different GDP per capita, and lifestyle. Price of oil, which influences the oil consumption for automobile, differs among countries. Unbalanced development between rural areas and urban areas is often observed in developing countries. Per capita energy consumption in urban areas is much higher than in rural areas in the same country. Urbanization causes a significant energy consumption increase.

Urban structure influences energy consumption within a country. Size, the density of land use, population density and a city's transportation system are controlling factors. The zoning of land uses is also an important factor. A discrete zoning system keeps residential areas far from businesses and industrial areas in a large city. The aim of such separation is to guarantee a comfortable environment in residential areas. However, such a zoning system causes long trips for residents who must commute between zones. Mixed-use zoning or a compact multi-nuclei city, as discussed later, would reduce commutation distance and carbon dioxide emissions from transportation.

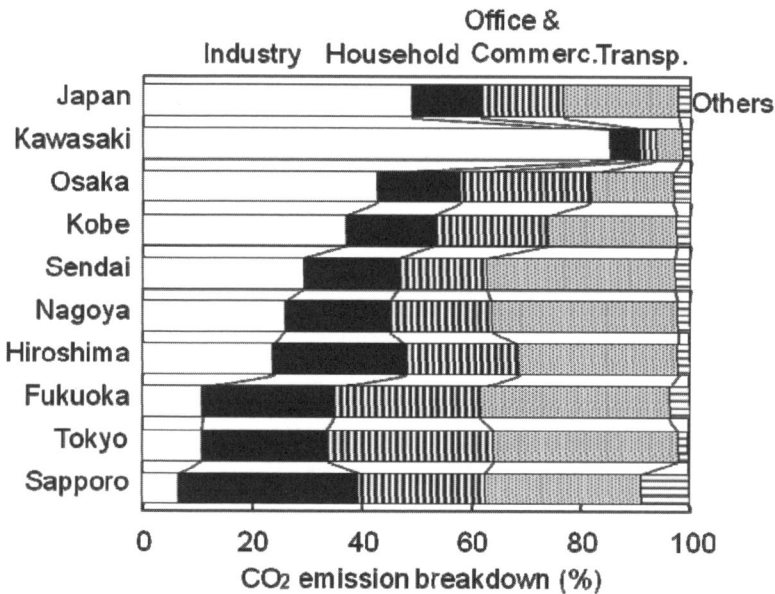

Fig. 10.5. Breakdown of carbon dioxide emissions by sector in large Japanese cities.

Profile of carbon dioxide emission in each city is diverse (Fig. 10.5). The contribution of each sector reflects the characteristics of the city. Kawasaki city is one of Japan's most industrial cities whereas Tokyo and Sapporo are typical commercial and service industry-based cities. Shares of industrial carbon dioxide emission differ very much among these cities. Each of these large cities plans a carbon dioxide emission reduction strategy, but the scope should be different. For example, household and commercial/office sectors are a substantial part of Tokyo's pattern of carbon dioxide emissions. The Tokyo Metropolitan Government has made a strategic plan called the "Tokyo Climate Change Strategy" to reduce carbon dioxide mainly from offices, households and transportation [3].

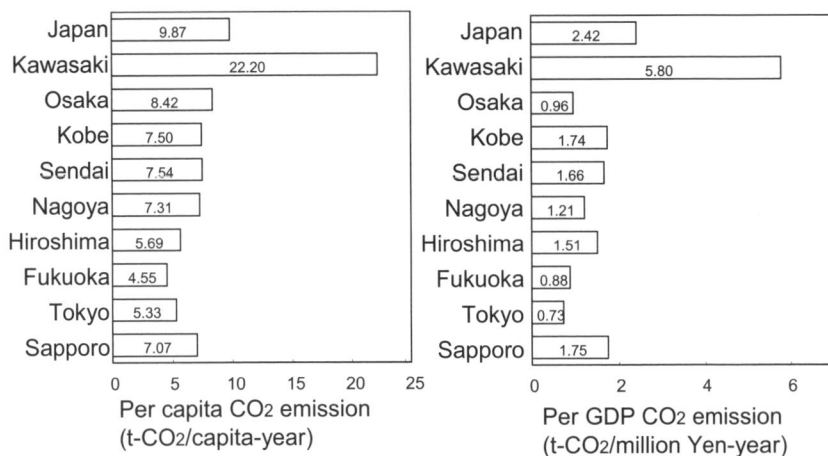

Fig. 10.6. Per person and per GDP carbon dioxide emission in large Japanese cities.

Differences among cities are not only due to the differences in sectors. Fig. 10.6 shows that the per person emission of carbon dioxide in Kawasaki city is far greater than in other large cities. Per GDP emission in Kawasaki is also much higher than in Tokyo. These high values do not mean that the Kawasaki is an energy-wasting city. The high amount of industrial activity in this city makes per capita and per GDP carbon dioxide emission much higher than the Japanese average. Other large cities show lower per capita and per GDP emission than the average value in Japan. This example shows that a simple comparison or ranking of the cities based on such values does not demonstrate energy or carbon dioxide efficiency for the urban area.

Fig. 10.7 shows per capita carbon dioxide emission in Beijing, Shanghai and Tokyo. The per capita average of carbon dioxide emissions in China is far smaller than in Japan, as is often mentioned. However, the emission totals of Beijing or Shanghai are already greater than Tokyo's though the estimated values contain uncertain factors derived from energy and population statistics in China. There are possible reasons for the high emission

rates of these Chinese cities.   The lifestyle in major cities such as Beijing
or Shanghai is close to that of Japan, and is much different from that in the
rural areas of China.   As coal is the major source of energy in China, the
same amount of energy consumption results in much higher carbon dioxide emission than in Japan.   The last point is that these Chinese cities
have higher levels of industrial activities than Tokyo.

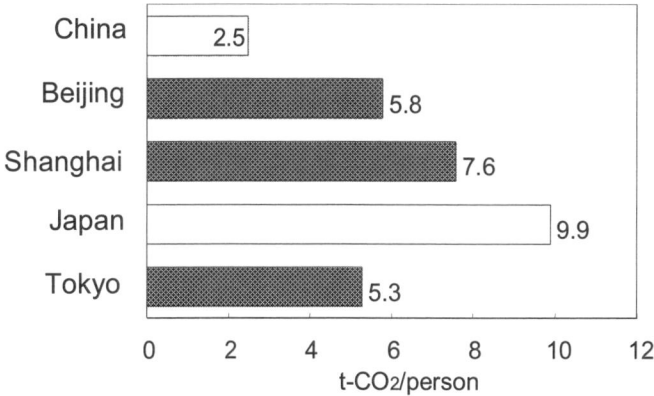

Fig. 10.7. Per capita carbon dioxide emission in Chinese and Japanese cities.
Chinese data is based on Dhakal [5]

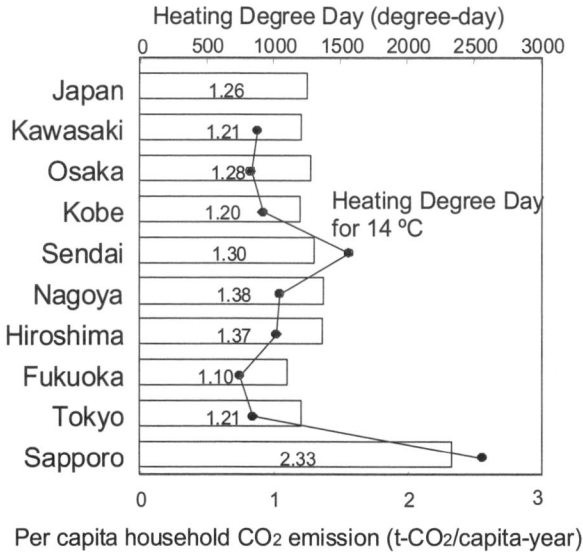

Fig. 10.8. Carbon dioxide emissions from households.

Carbon dioxide emissions from households in large cities (Fig. 10.8) does not differ very much except for Sapporo which is located in a much colder region than the other cities shown here. Such coldness is usually expressed in terms of the heating degree-day, which is defined as the 365 days summation of the temperature between daily average temperature and the designated standard temperature when the former is lower than the latter. Kawasaki City shows a pattern of household emission similar to that of other cities. Household energy consumption and the resulting carbon dioxide emissions basically depend on climate, lifestyle, floor area and family size.

Table 10.4. Trends of family and lifestyle change in household sectors in Japan.

a) House condition and family size				
Index	1973	1983	1993	2003
Family size (persons/household)	3.7	3.4	3.0	2.7
Floor area (square m/household)	77	86	92	96
Calculated floor area per person (square m/person)	20.8	25.3	30.7	35.6
Based on Housing and Land Survey of Japan.				

b) Lifestyle change				
Index	1970	1980	1990	2000
Energy consumption for hot water (10^9J/capita-year)	1.9	3.6	4.7	5.0
Energy consumption for "Power and other" category (10^9J/capita-year)	1.5	2.8	4.3	6.2
"Power and other" is energy consumption other than heating/cooling, hot water and kitchen uses [4].				

Table 10.4 a) shows that the family size in Japan has been decreasing since 1970. The average number in 2003 was only 2.7 persons. On the other hand, the floor area per household increased from 77 square meters in 1973 to 96 square meters in 2003. These two changes have increased per person floor area from 20.8 to 35.6 square meters/person during 1973 to 2003. Such an increase in floor area per person is good from the viewpoint of quality of life, because Japanese housing conditions are generally poor. However, one can easily understand that this results in the increase of energy consumption per person for lighting, space heating and cooling in the household sector. The other factor that has raised energy consumption is lifestyle change. Table 10.4 b) shows energy consumption per person for hot water and miscellaneous uses. Hot water is used for showers, which used to be less common. The "power and other" category includes

computers, televisions, and other appliances. As of 2000, this category had grown to more than four times the value of 1970, and now this category is too large to just be called "other."

Table 10.5 Trends of office/commercial energy consumption in Japan

a) Floor area increase					
Building type	1970	1980	1990	2000	2005
Total office and commercial (million square m)	560	936	1,286	1,656	1764
Office (million square m)	120	204	313	435	460
Department and retail store (million square m)	132	223	299	406	435
Data source: EDMC Handbook of Energy & Economic Statistics in Japan [4]					
b) Energy consumption					
Purpose	1970	1980	1990	2000	2005
Total (MJ/square m)	1222	1165	1,188	1,189	1,168
Cooling (MJ/square m)	43	53	96	131	130
Heating (MJ/square m)	520	446	311	269	244
Power and others (MJ/square m)	168	279	405	473	503
Data source: EDMC Handbook of Energy & Economic Statistics in Japan [4]					

Table 10.5 describes energy consumption in the office and commercial sectors. This category includes energy consumption in offices (19%), department and retail stores (22%), restaurants (9%), schools (7%), hotels (11%), hospitals (12%) and others (percentages shown in parentheses are the approximate share of energy consumption in this category in 2005 [4].) The table shows that energy consumption per floor area did not increase. Energy consumption by office machines has increased significantly, but heating energy consumption has been significantly reduced. The main factor that raises total energy consumption in the office/commercial sector is the increase of floor area, which has increased at a very high rate, as shown in Table 10.5a. Carbon dioxide emissions from this sector are difficult to be controlled unless the floor area for office use is controlled.

For energy consumption in houses or other buildings, energy efficiency is the governing factor. The improvement of insulation capacity, the type of fuel used, the efficiency of the heating or cooling machine, the type of light and home electric appliances all make a difference in reducing carbon dioxide emissions.

Table 10.6 indicates transportation energy use. Energy consumption in this sector varies greatly depending on the traffic mode. Private cars consume ten times as much energy as rail for the same amount of passenger traffic. Personal car trips have increased more than three times from 1970 to 2000. Because of these reasons, energy consumption of private car was 2.08 x 10^{12} MJ whereas rail consumed only 0.08 x 10^{12} MJ in 2000.

Table 10.6 Energy consumption of various modes of passenger and freight transportation in Japan.

a) Unit energy consumption for each traffic mode.			
Passenger traffic (MJ/person-km)		Freight traffic (MJ/t-km)	
Private Car	2.46	Truck	3.48
Bus	0.66	Rail	0.26
Rail	0.21	Sea	0.98
Air	1.98	Air	21.84

Value of 2005 [4]

b) Energy consumption

Index	1970	1980	1990	2000	2005
Passenger traffic					
Total Transportation (billion person-km)	703	891	1300	1,420	1,411
Personal Car (billion person-km)	278	414	727	852	833
Rail (billion person-km)	289	315	388	384	391
Total energy consumption (10^{12}MJ)	0.67	1.24	1.85	2.43	2.47
Personal Car	0.43	0.94	1.49	2.08	2.09
Rail	0.06	0.06	0.08	0.08	0.08
Freight traffic					
Total Transportation (billion ton-km)	352	442	547	578	570
Truck	137	182	274	313	335
Sea	151	222	245	242	212
Total energy consumption (10^{12}MJ)	0.77	1.06	1.26	1.37	1.34
Truck	0.54	0.79	1.08	1.12	1.10
Sea	0.18	0.24	0.15	0.22	0.21

Data source: EDMC Handbook of Energy & Economic Statistics in Japan [4]

Controlling the use of private automobiles is key in the control of carbon dioxide emissions by the transportation sector. Such private vehicles bring flexible mobility to citizens and they have enabled the development of new residential areas in many countries. However, they in turn cause environmental problems such as air pollution and carbon dioxide emission. A shift in transportation mode from private car to public transportation, called modal shift, can significantly reduce carbon dioxide emissions, but public transportation system such as railways require a densely populated or compact urban structure.

10.4. Control of GHG emissions from urban activities

Three different types of countermeasures can be implemented in urban areas to reduce GHG emission. These are: (1) carbon dioxide reduction in the energy and heat supply, (2) carbon dioxide reduction through energy consumption saving, and (3) change of urban structure. These three aspects should be combined together in order to achieve a substantial reduction. Integration of two or three of them is necessary to achieve a significant reduction in carbon dioxide emissions.

10.4.1. Carbon dioxide reduction in energy and heat supply

Power utility has been paying effort of lowering carbon intensity of its generated electricity. The reduction in carbon intensity owes to the selection of fuel source and energy conversion efficiency. Among the fossil fuels, coal produces the largest amount of carbon dioxide followed by oil and then natural gas. Natural gas is the cleanest fossil fuel from the viewpoint of carbon dioxide emissions.

Carbon intensity of electricity varies with the time of day and the season. It should be noted that the average carbon intensity of electricity depends on demand. Meeting peak demand of electricity requires fossil fuel power plants that are flexible to the demand change with time. On the other hand, hydropower and nuclear power are major sources in the off-peak time. Therefore, carbon intensity in the peak time in summer is the highest throughout the year. Averaging the electricity demand in urban areas throughout the year can contribute to lowering the carbon intensity of power generation.

In addition to that, distributed electricity generation systems have the potential to reduce carbon dioxide in urban areas. One example is cogeneration, in which electricity and heat are supplied simultaneously from city gas or oil. As conversion efficiency of fuel to electricity is usually 40 to 50%, more than half of fuel energy is wasted in power plants if it is not used. It should be noted that the implementation of cogeneration does not always reduce total carbon dioxide emissions. If the heat is not used properly, carbon dioxide emissions from providing electricity can be higher than that of grid power which partly comes from renewable or atomic energy.

There are heat sources in urban areas that are not well utilized. Heat from solid waste incineration and heat from domestic sewage are potential heat sources. District heating systems are necessary to utilize these sources. This means that heat demand with high density is necessary to implement this system. However, many of the solid waste incineration plants and wastewater treatment plants are not located in the central business district. This is a problem of mismatched infrastructure locations.

10.4.2. Carbon dioxide reduction in energy saving

There are various countermeasures to reduce energy consumption in the non-industrial sector. These include building energy use/construction, energy-saving buildings, long-life buildings, automobile innovation (hybrid vehicle, fuel cell vehicle), transportation system (traffic demand management) and district heating in high density area with urban heat recovery. Many technologies developed in the building and transportation sectors should be implemented, along with proper management of energy demand. For example, automobile innovation and reduction of traffic demand should be done at the same time.

The mix of countermeasures implemented should depend on the type of city. The methods used in megacities are not necessarily effective in small cities. In large cities, efficient apartment houses can be promoted. Detached houses, which have a large roof area, are suitable for photovoltaic cell installation. Cogeneration and district heating/cooling systems can be applied only to an appropriate area. Modal shift from automobile to train should be promoted in cities where the share of the rail is not high. In region-core city, typically with a population of around several hundred thousand, railway systems can be implemented depending on the distribution of traffic demand. LRT (Light Rail Transit) is one option for public transportation systems. In smaller cities or villages, the use of public transportation is difficult from the viewpoint of business profit and even from the viewpoint of energy saving. Effective use of the automobile is the only method to reduce carbon dioxide emission in those settings.

The utilization of biomass from forests and agriculture can make rural areas more environmentally friendly, though the levels of emissions from agricultural and forest industries are not currently high. Promoting the utilization of biomass together with the development of small towns is promising.

10.4.3. Change of urban structure

The structure of the city, especially the land use plan, influences the energy consumption level of the whole city. Usual urban planning has a zoning plan which limits the uses of land to preserve a good living environment in residential areas (Fig. 10.9). However, the separation of residential areas from business areas in large cities causes long commute times for people. This causes higher energy consumption in the transportation sector. Mixed land use offers a possibility to shorten the commutation distance. A disadvantage of this plan is the possible deterioration of the urban environment because of uncontrolled land use.

Compact cities have been promoted as a way to decrease urban environmental burden. A multi-nuclei structure becomes necessary in the case of a large city, as shown in the figure. The compact city has a structure consisting of the CBD (central business district), a commercial zone and a residential zone with shorter travel distances among these zones. The presence of high density in the CBD enables the implementation of a district heating and cooling system. Such a compact city can create open space in suburban areas and among nuclei cities. Vegetation can be recovered in such open space.

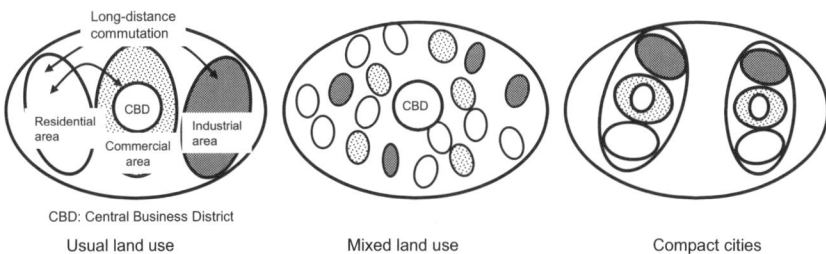

Fig. 10.9 Land use and the compact city

10.5. Direct and induced carbon dioxide emission

As mentioned earlier in section 3.2, a service-oriented city like Tokyo or Sapporo apparently emits a low amount of carbon dioxide per person, whereas an industrial city like Kawasaki emits a far higher level of carbon dioxide.   However, service-oriented cities consume a large amount of goods that are produced elsewhere and emit carbon dioxide and other pollutants (Fig. 10.10). Such environmental loading is called induced loading or embodied loading in contrast with direct loading. For example, carbon dioxide from gasoline combustion from cars running in a city is direct carbon dioxide emission, whereas carbon dioxide emission during the manufacturing of the car is induced carbon dioxide emission. This induced carbon dioxide emission is not taken into account in usual comparisons of carbon dioxide emission among cities.

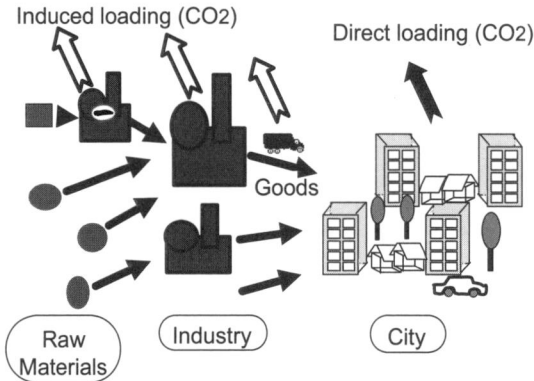

Fig. 10.10. Direct and induced emissions by urban activity.

Such induced environmental loading can be estimated at the city scale by analyzing the regional input/output table. Fig. 10.11 shows the direct and induced carbon dioxide emission in selected cities. Only 30% of the carbon dioxide emission arising from the activity of Tokyo is counted as direct emission. The remaining 70% of carbon dioxide is emitted in other cities in order to support the activities of Tokyo. Shanghai is a service city which also shows a low percentage of direct emission. Fukuoka and Kita-kyushu are located within the same prefecture, but their function is different. Kita-kyushu is an industrial city while Fukuoka is a commercial city. There is a strong contrast between the percentages of direct and induced emission between these two cities. An analysis of induced carbon dioxide reveals that there is a large degree of hidden environmental loading in commercial or service cities. Such loading should be taken into account when the environmental loading from urban activities is to be reduced.

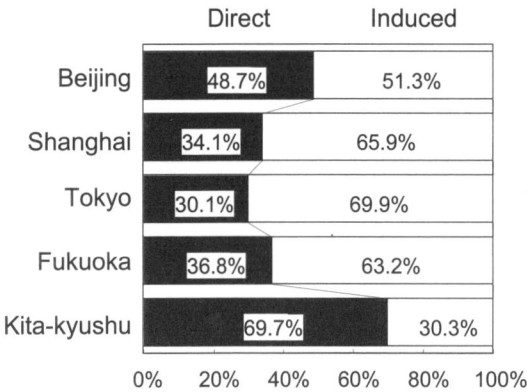

Fig. 10.11. Direct and induced carbon dioxide emission by several cities (Data source: Beijing and shanghai [6], Tokyo [7], Fukuoka and Kita-kyushu [8].)

10.6. Summary

Urban activities and climate change have an interactive relationship. Climate change would impact urban activities through water resource shortage problems and rising sea levels. The contribution of urban activities to climate change is very significant. Proper management of urban activities is one of the key issues in a strategy to reduce carbon dioxide and other greenhouse gases. Improvement of urban activity, especially reformation of city structure, usually takes long periods of time. Action should be taken now to achieve the long-term target of climate change control by the middle of the 21st century.

References

[1] United Nations Population Division World Urbanization Prospects: The 2005 Revision Population Database. Accessible at http://esa.un.org/unup/

[2] Intergovernmental Panel on Climate Change (2007), 4th Assessment Report (AR4) of Working Group 2, Climate Change Impacts, adaptation and vulnerability.

[3] Tokyo Metropolitan Government (2007), "Tokyo Climate Change Strategy", English version is available at
https://www2.kankyo.metro.tokyo.jp/kouhou/env/english/index.html

[4] The Energy Conservation Center, EDMC Handbook of Energy & Economic Statistics in Japan, 2007 version and other year versions.

[5] Dhakal, S. (2004), Urban Energy Use and Greenhouse Gas Emissions in Asian Mega-Cities: Policies for a Sustainable Future, Institute for Global Environmental Strategies (IGES), Japan

[6] Shinji Kaneko, Hirofumi Nakayama, Libo Wuc (2003) "Comparative study on indirect energy demand, supply and corresponding CO2 emissions of Asian Mega-cities", Proc. International Workshop on Policy Integration Towards Sustainable Urban Energy Use for Cities in Asia, Institute for Global Environmental Strategies.

[7] Yoshida, Y., Ishitani, H. and Matsuhasi, R. (1998): Classifying CO2 emission from the viewpoint of LCA by reflecting the influence of regional activities, Journal of the Japan Institute of Energy, Vol. 77, No.11, 1054-1061 (in Japanese)

[8] Kanagawa, T., Kato, E. and Imura, H. (1992) Application of the input-output model to the analysis of urban energy consumption, Environmental Systems Research, Vol.20, 242-251 (in Japanese).

11. Japanese Strategies for Global Warming Issues and Energy Conservation in Buildings

Yuzo Sakamoto

Department of Architecture, The University of Tokyo
7-3-1, Hongo, Bunkyo-ku, Tokyo, 113-8656, Japan

11.1 Global Warming Issues and CO_2 Emission

Global warming, which was called climate change in Chapter 10, is caused by anthropogenic actions. Gases causing global warming are called green house gases (GHG), and the most significant GHG is carbon dioxide (CO_2). In fact, enormous amounts of CO_2 have been emitted since the 19th century from the activities of human beings, for instance, industry, transportation and the utilization of buildings. Most scientists believe that the increase of CO_2 concentration in the atmosphere causes global warming and that global warming and climate change will bring human beings many disasters and widespread damage, as was shown in Chapter 10. These problems are called "Global Warming Issues."

A main topic regarding these problems is the global reduction of GHG, CO_2 in particular, emitted into the atmosphere. Fig. 11.1 shows the share of CO_2 emissions by country. The U.S.A. has the highest share of emissions, followed by China. Japan's position is fourth in this respect. In December of 1997, the Japanese Government agreed to the Kyoto Protocol, an international treaty proposed by the Intergovernmental Panel on Climate Change of the United Nations, in order to define reduction percentages of CO_2 emissions country by country. Consequently, the Japanese Diet ratified the Protocol in 2002. Thus, in Japan, energy conservation became very important, as CO_2 emissions are mostly caused by energy consumption. However, even if Japan didn't agree to the Kyoto Protocol, energy conservation is important for Japan because of a scarcity of energy resources.

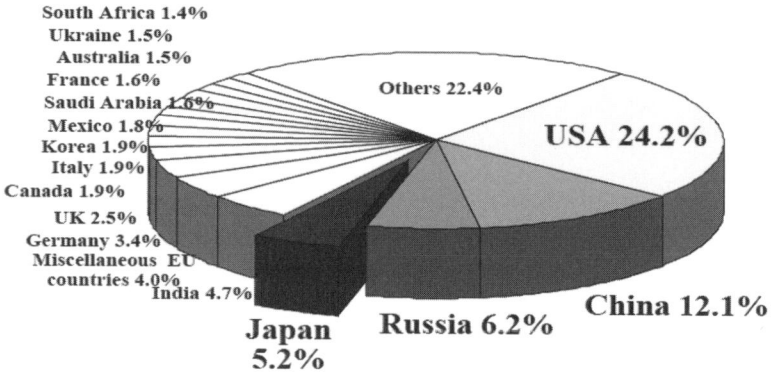

Fig. 11.1. Share of CO_2 emissions in 2000 (divided by country) [1]

Fig. 11.2 shows the trend of annual CO_2 emissions in Japan from 1990 to 2003. CO_2 emissions are gradually increasing in Japan. According to the Kyoto Protocol, Japan has promised to reduce emissions by 6% of the 1990's emission level, which amounts to 1.1 billion tons in CO_2 weight and which is shown by the dotted line in Fig. 11.2. Nevertheless, the recent emissions of Japan are over the emission level of the 1990s. Fig. 11.3 explains the reason why Japan's CO_2 emissions are gradually increasing. The emissions in the housing and building sectors show consistent increase from 1990. Consequently the recent incremental rate in both sectors exceeds 30%, and such a large incremental rate is the cause that the recent emissions of all sectors are over the emission level of the 1990s.

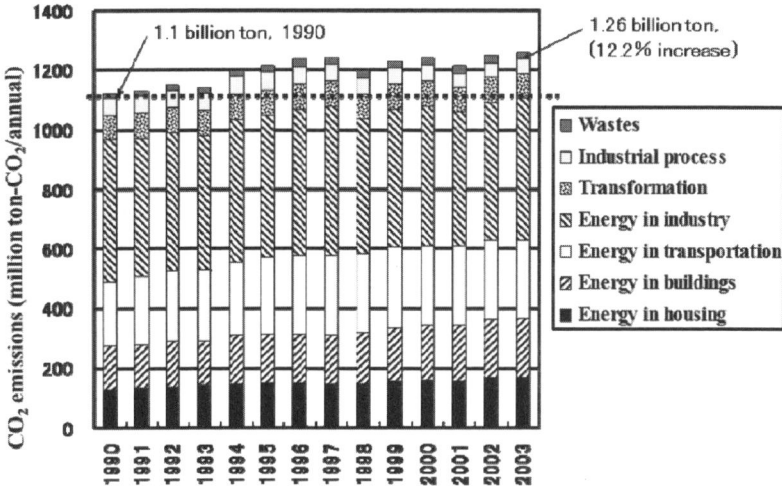

Fig. 11.2. Trend of annual CO_2 emissions in Japan [2]

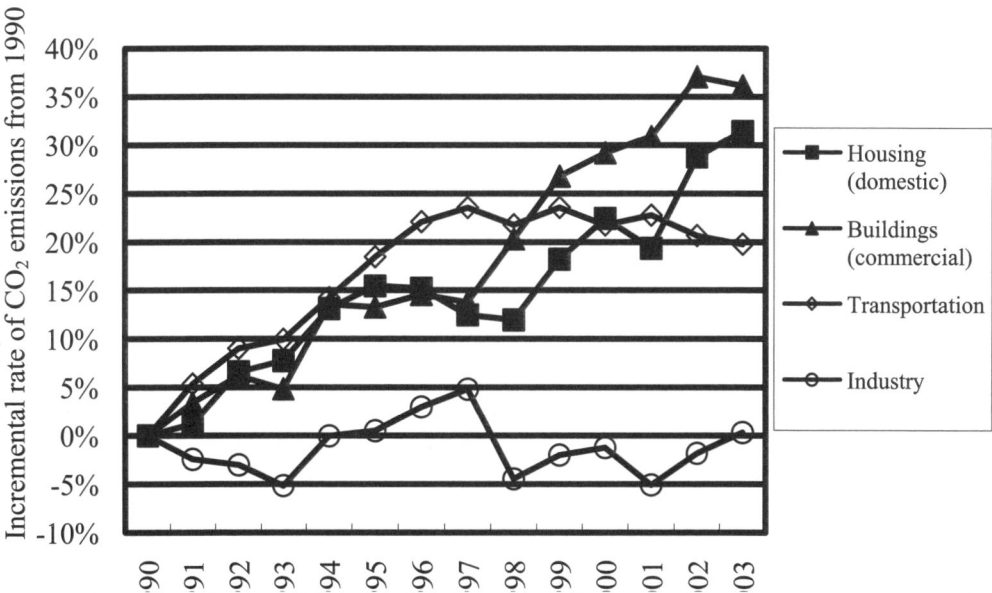

Fig. 11.3. Increased rate of CO_2 emissions by sector [2]

Thus saving energy in housing and buildings has become more important for Japan and its goal to adhere to the Kyoto Protocol. In that regard, the Japanese Government has made action plans to reduce the CO_2 emitted in the housing and buildings sectors. The amount of the planned reduction is 97.2 million tons of CO_2 compared to "business as usual." One of the powerful methods in the action plans is the dissemination of energy conservation standards for housing and buildings, as presented in the Table 11.1. Through the dissemination process, the Japanese Government expects to stimulate a 34 million ton reduction of CO_2 emissions.

Table 11.1 Ways of saving energy in buildings and housing, announced by the Japanese Government in 2005

Ways of saving energy	CO_2 reduction (10^6ton)
Dissemination of "energy conservation standards"	34.0
Improvement of electric appliance efficiency	38.8
Dissemination of BEMS[*1]and promotion of ESCO[*2]	11.2
Improvement of combustion efficiency	4.9
Dissemination of energy-efficient illumination	3.4
Dissemination of heat pump hot water boiler	3.4
Reduction of standby power	1.5
Total	97.2

[*1] Building energy manage systems
[*2] Energy saving companies

In Japan, energy conservation standards have been established on the basis of "The Energy Conservation Law." Fig. 11.4 illustrates the framework of countermeasures outlined in The Energy Conservation Law. The Energy Conservation Law is composed of four sector-based countermeasures: for factories, for buildings and housing, for transportation and for machinery. However, these countermeasures are not mandatory in Japan. Instead, they are recommendations. The buildings and housing energy conservation standards are composed of two different standards. One is the standard for housing, which requires the reduction of thermal load. The other is the standard for buildings, which includes reduction of thermal load and improvement of equipment system efficiency. In particular, the reduction of thermal load is very important in housing. Namely, the standards require fundamental technologies for energy conservation in housing and buildings.

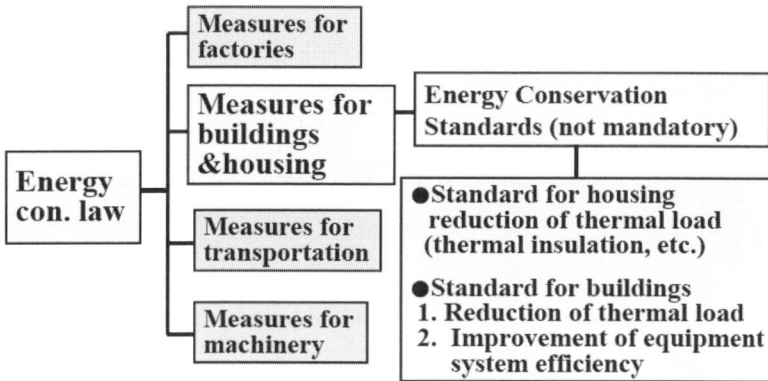

Fig. 11.4. Framework of the Energy Conservation Law

11.2 Energy Conservation in Housing

11.2.1 Envelope Design

Envelope design for high thermal performance is a fundamental technology to create a comfortable room environment using less energy. It means a high level of insulation, high air-tightness, proper moisture control and appropriate solar radiation control in walls, roofs, floors, windows and other components. If we design an excellent envelope for rooms and buildings as a whole, we can obtain both energy conservation and comfortable room temperatures.

Fig. 11.5 presents a comparison of energy consumed in international housing. Japan has a much lower value than European countries. In particular, heating consumption in Japan is much less than in European countries and the U.S.A. The reason is that heating equipment is poor in most areas of Japan. In areas besides Hokkaido, not all rooms are heated, even in mid-winter, and portable kerosene heaters are usually used instead. The kerosene heater is a very popular type of heating equipment, as is the heat-pump air-conditioner. Approximately 5 million heaters are sold in Japan every year [3].

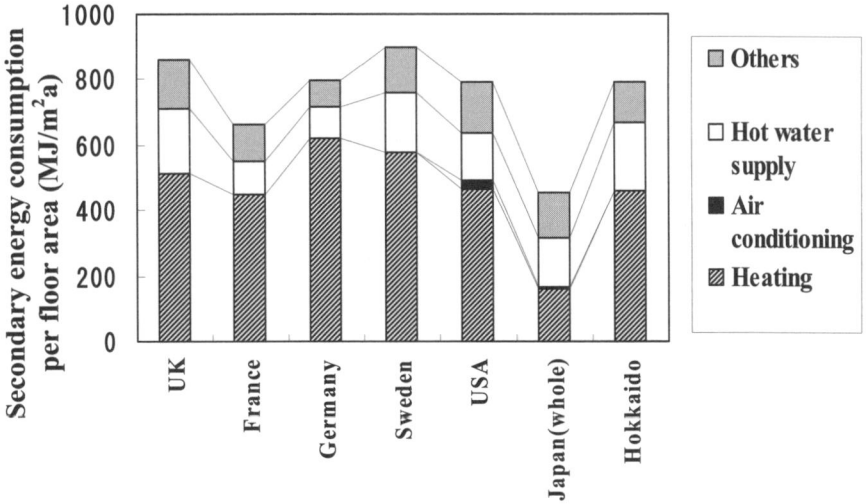

Fig. 11.5. Comparison of energy consumption in international housing [4]

Fig. 11.6 shows traditional Japanese style housing. These types of homes are found sometimes all over Japan, but percentage of them is very small. Namely, the homes have large windows and many doors along with long eaves, which are characteristic of Japanese housing. The homes have

- ●**Wood frame structure**
- ●**Large openings(windows) and deep eaves**
- ●**Continuity between indoors and outdoors**

Fig. 11.6. Traditional Japanese style housing

a sense of continuity between indoors and outdoors. The indoor climate in the homes is very cold in winter because of enormous heat loss. However, during summer nights, the indoor climate is cool. Thus, we have to improve the feeling on indoor coldness in winter, but keep the sense of cool comfort during the other seasons.

Fig. 11.7 shows one concept called "closing and opening in housing." "Closing" refers to insulation and the use of a sunshade, whereas "opening" refers to cross ventilation by opening windows. We need "closing" both during heating time in winter and when using air-conditioning in summer. On the other hand, Japanese people tend to prefer "opening" if the outdoor air is comfortable during spring, autumn and summer nights. So we can take in outdoor air through the large windows during these times. In conclusion, it can be stated that harmony between climate and environmental performance is important in Japan. Thus we need both "closing" and "opening." According to the energy conservation standards for housing, thermal insulation and an airtight envelope are regarded as indispensable elements for "closing." [5]

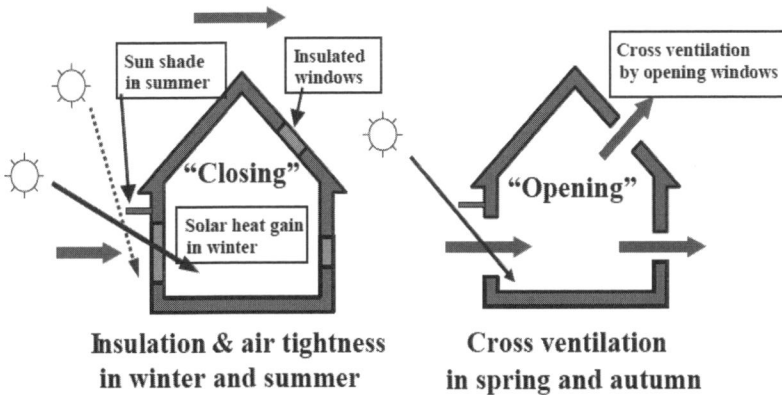

Insulation & air tightness in winter and summer   **Cross ventilation in spring and autumn**

Fig. 11.7. "Closing" and "opening" in housings

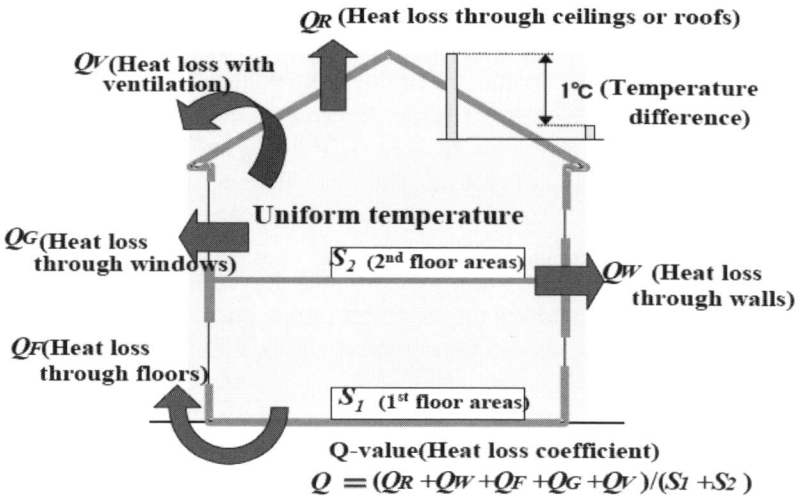

Fig. 11.8. Definition of Q-value, used as an indicator for thermal insulation

Q-value, an indicator for thermal insulation, is introduced and defined by the equation shown in Fig. 11.8. According to this equation, the Q-value is calculated by dividing total heat loss by the total floor area, where total heat loss is composed of heat loss through building components and through ventilation. Thus the Q-value is very important and is a very sensitive indicator of thermal insulation performance, as shown in Fig. 11.9. The figure shows that a smaller Q-value corresponds to a smaller heating load in all cities. Therefore, in the energy conservation standards for housing, the criteria related to envelope insulation have been defined using Q-values. Fig. 11.9 also illustrates the criteria for the Q-value in the three standards and shows the differences of heating load among the different standards. Fig. 11.10 is another example, showing the influence of Q-value on the temperature difference between indoors and outdoors. The temperature difference is inversely proportional to the Q-value. So the temperature difference approaches 12K in the case of much solar heat gain, where the Q-value is $1.6 W/(m^2 \cdot K)$.

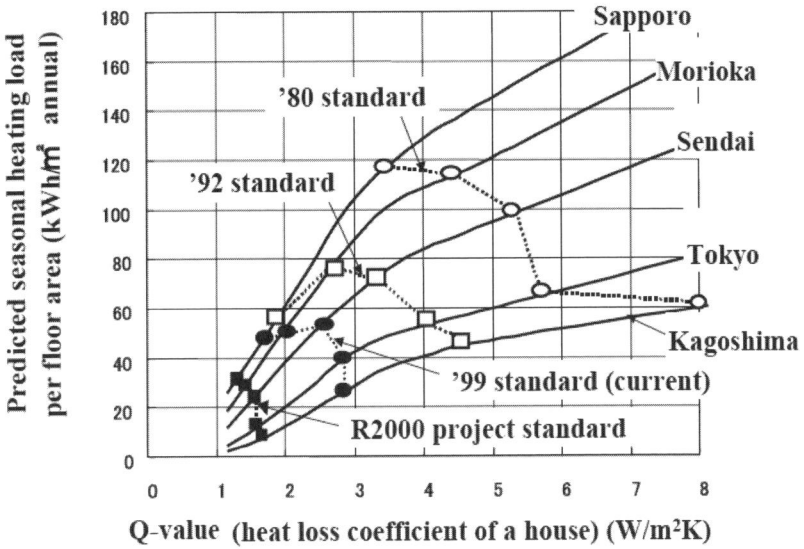

Fig. 11.9. Effect of Q-value on heating load and criteria for Q-value in the energy conservation standards

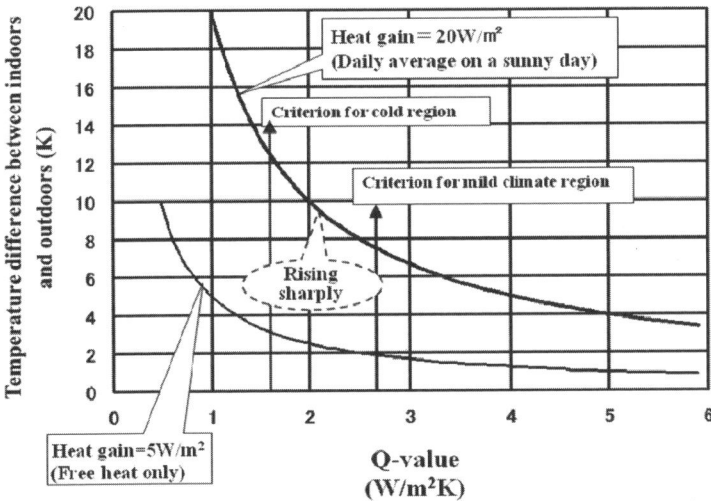

Fig. 11.10. Influence of Q-value on the temperature difference between indoors and outdoors.

The energy conservation standards have another indicator, μ-value. This indicator is related to solar heat gain as shown in Fig. 11.11 and is used as a sunshade parameter. Thus, if the μ-value is low, room temperature approaches outdoor temperature because the house does not get much solar heat. The μ-value is defined as $I / (JoS)$, where I is heat gain in an actual house and where JoS is solar heat gain on imaginary floors without a house.

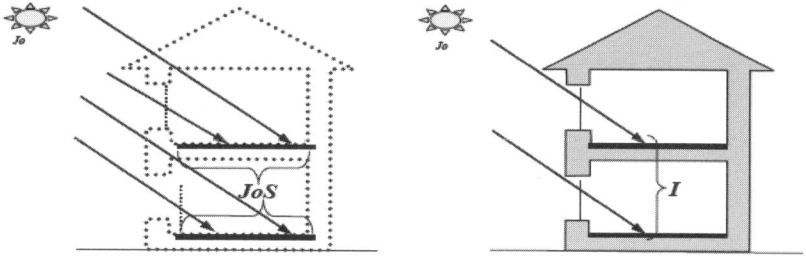

Solar heat gain coefficient, $\mu = I / (JoS)$,
where I=solar heat gain, S=total floor areas and
Jo= solar radiation on the horizontal surface.

Fig. 11.11. Definition of μ-value, used as a sunshade indicator

1. Fill insulation in post & beam construction   **2. Exterior insulation in post & beam construction**

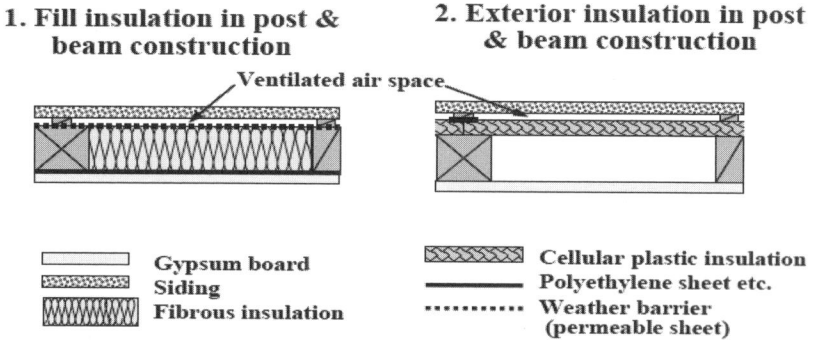

Ventilated air space

Gypsum board	
Siding	
Fibrous insulation	

Cellular plastic insulation
Polyethylene sheet etc.
Weather barrier (permeable sheet)

Fig.11.12. Typical examples of insulated wall assemblies in wood frame structures

Fig. 11.12 shows typical examples of insulated wall assemblies in wood frame structures in Japan. Residential buildings are mostly built using wood frame structures in Japan. Recently, as envelope insulation has been recognized as very important, there are diverse insulation specifications in Japan. In Fig. 11.12, the first figure shows the most typical insulation wall in use. This is called "fill insulation" used in post-and-beam- construction. The second figure presents "exterior insulation" used in post-and-beam-construction. This second type has become common recently.

Window insulation is as important as wall insulation. The first picture in Fig. 11.13 is an example of bad window insulation showing an old type window composed of a single pane and aluminum frame. It results in a great deal of heat loss and in vapor condensation that causes mold and mites on the surface. These kinds of windows are found in most Japanese houses. The second picture is an example of good window insulation, which is composed of low-emissivity-double-pane glass and a plastic frame It results in less heat loss. In the case of this window, as the temperature on the surface of the glass grows closer to the room air temperature, vapor condensation does not take place.

An old type window
= Single pane + Aluminum frame
   →Much heat loss
   →Much condensation (non
     sanitary because of mold
     and mites)

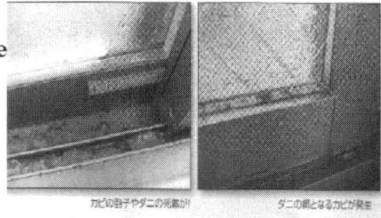

A good insulation window
= Low-emissivity-double-pane +
  Plastic frame
   →Less heat loss
   →Sanitary and healthy rooms

Fig. 11.13. Comparison between an old type and an insulated type of window

11.2.2 Saving Energy Using Heat Pumps in Housing

Heat pumps are a fundamental technology to produce various kinds of heat energy efficiently. They are used for heating, air-conditioning, supplying hot water, refrigerating and so on in housing and other buildings. Fig. 11.14 shows that heat demand makes up 64% of the total energy consumption in housing. A major part of the heat demand consists of supplying hot water.

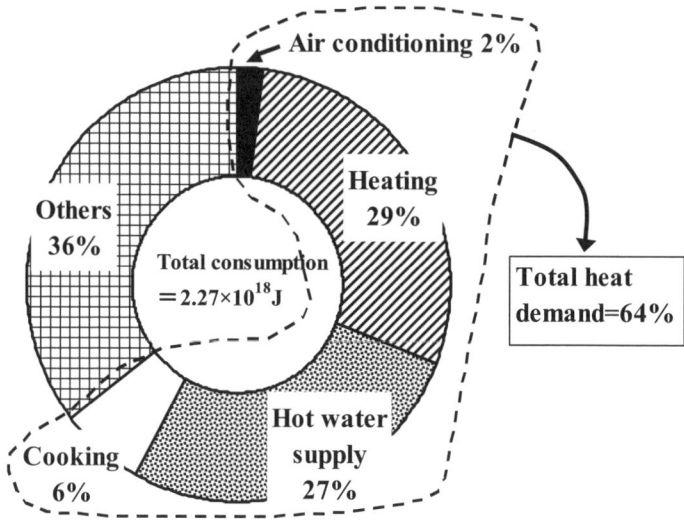

Fig. 11.14. Details of domestic energy consumption in Japan [6]

A heat pump is a machine that takes heat energy from a heat source and supplies that heat energy to a heat sink, as shown on the left of Fig. 11.15. The heat pumps works through mechanical force, which is usually provided by electric motors. A theoretical equation for the COP(coefficient of performance) of an ideal heat pump is formulated in terms of thermodynamics, as shown below in Fig. 11.15. COP is an indicator that shows the energy efficiency of a heat pump, and is defined as the ratio of output heat to the electric energy input. Three curves in the graph on the right indicate the COPs derived from that equation. The maximum values of COP among existent air-conditioners and hot-water-suppliers are also plotted on the graph. In addition, a dashed line on the graph presents the COP that is equivalent to 100% energy efficiency in a combustion boiler. The graph shows that the energy-efficiency of a heat pump is much higher than that of a combustion boiler in Japan.

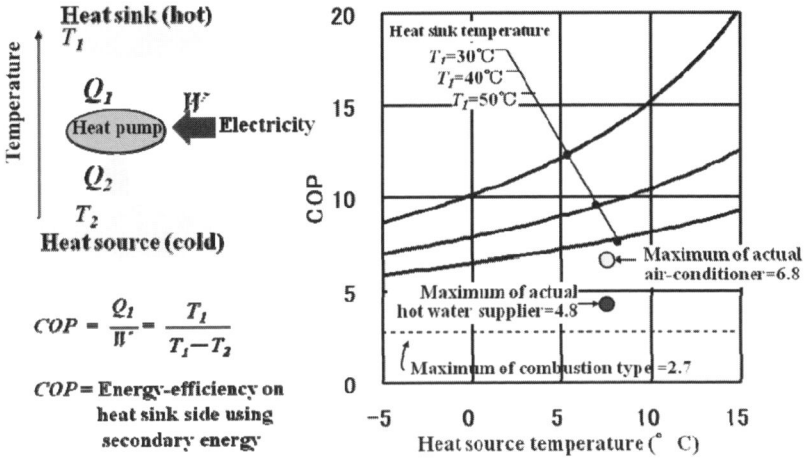

Fig. 11.15. Theoretical energy-efficiency of an ideal heat pump

The above is illustrated by Fig. 11.16. If we have 100% primary energy in fossil fuel, 37% in electric energy is obtained at Japanese thermal power stations, because the average efficiency of a generator is 37% according to the Energy Conservation Law. Then, using an old type of electric heater, we can get 37% in hot water energy. However, if we choose a heat pump type of electric heater, such as the one named "eco-cute," we can get more than 111 in hot water energy. The reason is because the annual average COP of an "eco-cute" is larger than 3. On the other hand, if we use a usual combustion boiler, we can only get 78% in hot water energy. Thus, the "eco-cute" heater has a much higher performance supplying hot water than that of a usual combustion type of heater.

Consequently, the advantages of heat pumps for energy savings are shown as follows:

1. Heat pumps have higher energy efficiency than combustion type heaters given a mild winter climate, which is true for most of Japan.
2. Moreover, the warmer the winter climate is, the higher the energy efficiency is for the heat pumps.
3. Heat pumps are a mature technology, so they are not very expensive. The price is reasonable, compared with fuel cells, etc.
4. Various improvements to the heat pumps (smaller size, cold climate types, etc.) are expected in the future.

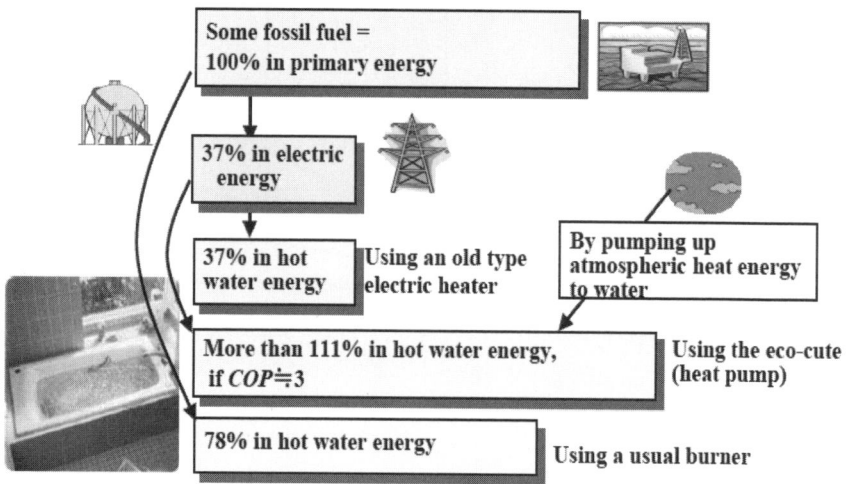

Fig. 11.16. Comparison of energy performance among three types of hot water suppliers

11.3 Energy Conservation in Non Residential Buildings

Non-residential buildings are classified into several categories. Fig. 11.17 shows energy consumption data of the five categories of non-residential building: offices, hospitals, schools, hotels and department stores. The total amount and details of energy consumption are different in each category. However, energy consumption for air conditioning is significant in all categories of buildings.

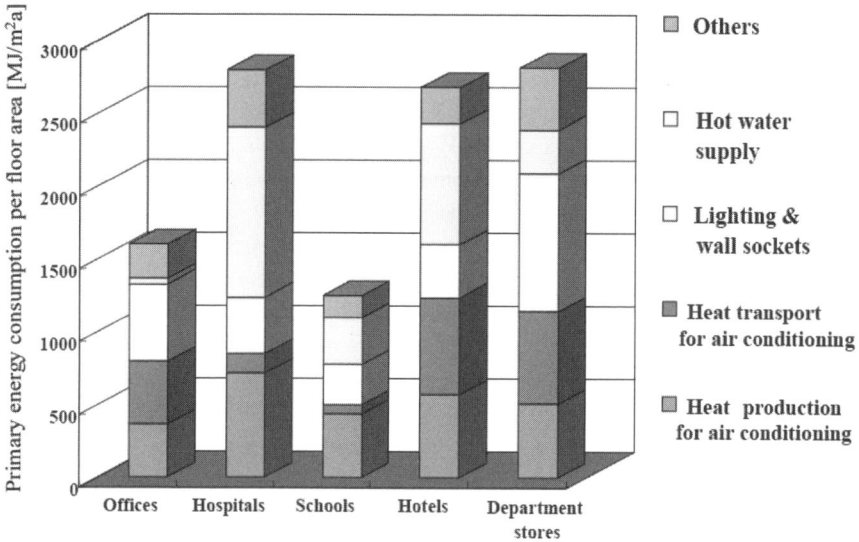

Fig. 11.17. Energy consumption in five categories of non-residential buildings [7]

Fig. 11.18 shows three strategies to save energy in buildings: envelope design, equipment system efficiency and energy management. Under the present energy conservation standards, there are independent criteria for envelope design and for equipment. For envelope design, PAL (perimeter

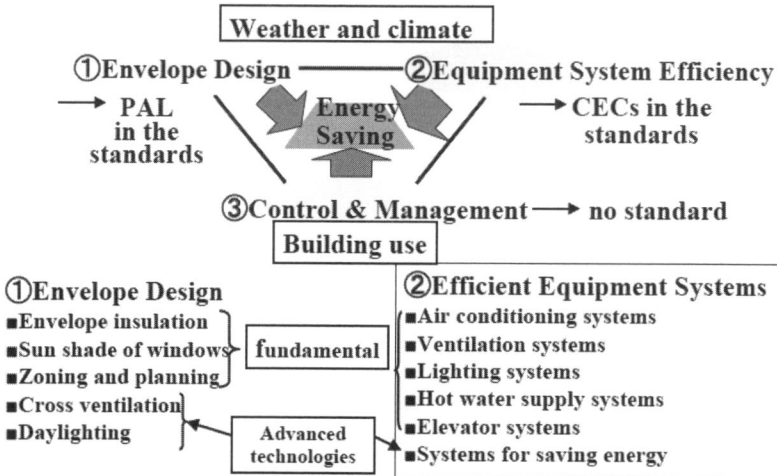

Fig. 11.18. Three strategies to save energy in buildings

annual load) is an indicator to show annual thermal load; for equipment design, another indicator, CEC (coefficient of energy consumption), is used.

PAL is used to evaluate envelope insulation, sunshade of windows and building plans. However, passive technologies like cross ventilation and daylighting are not evaluated by PAL. CEC is composed of five kinds of CEC: CEC/AC for air-conditioning systems, CEC/V for ventilation systems, CEC/L for lighting systems, CEC/HW for hot water supply systems and CEC/EV for elevator systems. These CECs are the indicators for energy efficiency of the respective pieces of equipment.

The definitions of PAL and CEC are shown in Fig. 11.19. PAL is the annual perimeter thermal load divided by the perimeter floor area. The perimeter is illustrated in the figure. The criteria for PAL and CEC's are listed in Table11.2.

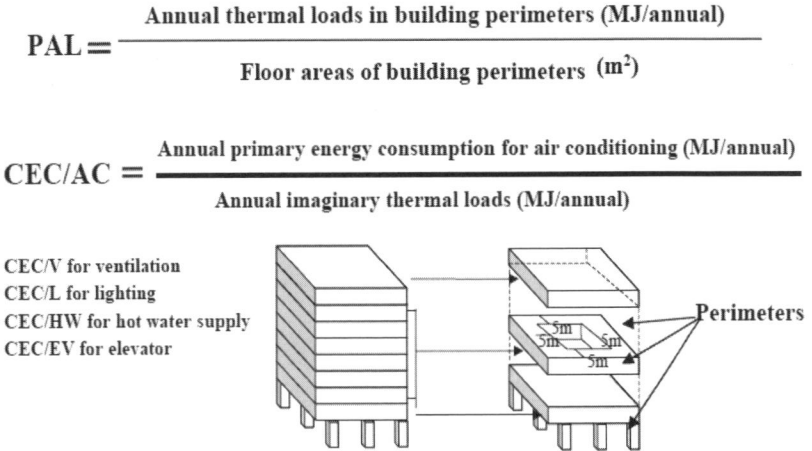

$$PAL = \frac{\text{Annual thermal loads in building perimeters (MJ/annual)}}{\text{Floor areas of building perimeters } (m^2)}$$

$$CEC/AC = \frac{\text{Annual primary energy consumption for air conditioning (MJ/annual)}}{\text{Annual imaginary thermal loads (MJ/annual)}}$$

CEC/V for ventilation
CEC/L for lighting
CEC/HW for hot water supply
CEC/EV for elevator

Perimeters

Fig. 11.19. Definitions of energy performance indicators (PAL and five CECs)

Table 11.2 Numerical criteria in the energy conservation standards

	Hotels	Hospitals	Shops	Offices	Schools	Restaurants	Theaters	Factories
PAL(MJ/m²a)	420	340	380	300	320	550	550	--
CEC/AC	2.5	2.5	1.7	1.5	1.5	2.2	2.2	--
CEC/V	1.0	1.0	0.9	1.0	0.8	1.5	1.0	--
CEC/L	1.0	1.0	1.0	1.0	1.0	1.0	1.0	1.0
CEC/HW	1.5~1.9, defined by hot water volume ÷ pipe length							
CEC/EV	1.0	--	--	1.0	--	--	--	--

11.4 Advanced Technologies and Actual Energy-Efficient Buildings

There are three categories of advanced technologies for saving energy:
1. Renewable energy: solar heat, photovoltaic power, wind power, biomass fuel energy, etc.
2. Utilization of wasted heat sources: cogeneration, sewage water, river water, ground water, etc.
3. Hybrid systems using passive methods: passive solar, cross-ventilation, daylight, etc.

However, these are not evaluated in the energy standards and are referred to as advanced technologies in this chapter. Fig. 11.20 illustrates the passive methods that are used in hybrid systems. Cross-ventilation is useful in reducing air-conditioning load in spring and autumn season. Daylighting is also effective in reducing illumination energy in large window spaces.

Shown below are three examples of actual energy-efficient buildings. Fig. 11.21 shows the Itoman City Hall in Okinawa Prefecture. Sunshade louvers, on which photovoltaic panels have been placed, are used in this building. Fig. 11.22 shows Tokyo Gas' "Earthport Building" located in Yokohama. This building has a very large atrium that is located on the north side of the building. Thus the building obtains a lot of daylight and comfortable outdoor air during the proper time or season. The building is built using the concept of a hybrid system.

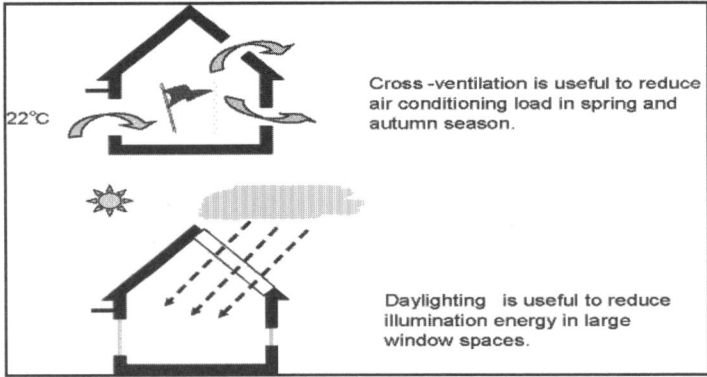

Cross -ventilation is useful to reduce air conditioning load in spring and autumn season.

22°C

Daylighting   is useful to reduce illumination energy in large window spaces.

Fig. 11.20. Passive methods to save energy

Envelope design : sun shade louvers
Renewable energy : photovoltaic power
Sun shade louvers using photovoltaic panels

Fig. 11.21. Itoman City Hall in Okinawa (an energy-efficient building in a tropical climate)

◆ **Passive methods : cross-ventilation and daylighting**

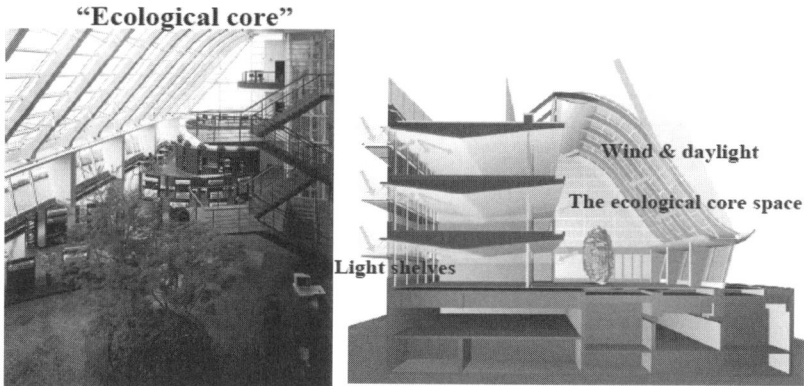

"Ecological core"

Fig.11.22. Tokyo Gas' "Earthport Building" (a hybrid system using cross-ventilation and day lighting)

Before retrofitted (built in 1961)
PAL=248MJ/m²a
CEC/AC=1.39

After retrofitted (retrofitted in 2000)
PAL=225MJ/m²a
CEC/AC=0.98

Fig. 11.23. "Oak Tokyo Building" (a building retrofitted for energy savings)

Finally, the Oak Tokyo Building is shown in Fig. 11.23 as an example of retrofitting. Actually, retrofitting is more important for Japan and its goal is to realize energy savings over the next few decades, as there are enormous buildings that are 30 - 40 times larger in floor area than many of the buildings being built currently. By retrofitting the Oak Tokyo Building, the PAL was improved from 248MJ/m²a to 225MJ/m²a. Similarly, the CEC/AC was reduced from 1.39 to 0.98.

References

[1] Oak Ridge National Laboratory (2000)
[2] Japan Center for Climate Change Actions (2005)
[3] Japan Industrial Association of Gas and Kerosene Appliances (2005)
[4] Institute for Building Energy Conservation, Japan (2002), " A Guidebook to the Energy Conservation Standards for Housing" (in Japanese)
[5] Y.Sakamoto (1999), Energy Conservation Standards for Residential Buildings, Revised for Japanese Next Generation, Proceedings of Fifth Canada-Japan Housing R&D Workshop, Charlottetown, 351-362,
[6] The Energy Conservation Center, Japan (2005)
[7] Institute for Building Energy Conservation (2005) " A Guidebook to the Energy Conservation Standards for Buildings" (in Japanese)

12. Management of Urban Heat Environment

Yasunobu Ashie

Designated Visiting Researcher Courses, The University of Tokyo
7-3-1, Hongo, Bunkyo-ku, Tokyo, 113-8656, Japan

12.1 The Urban Heat Island Mechanism and Its Effects

12.1.1 Urban heat islands

The urban heat island effect describes a phenomenon in which urban temperatures exceed those of areas outside the city, as shown in Figure 12.1. When contour lines of the temperatures are drawn on a horizontal plane, the resulting figure appears like an island in the middle of an ocean, hence the term "heat island." The urban heat island effect was first observed in London in the 19th century [1], and there have since been numerous observations of heat islands in cities around the globe. Urban heat islands are most prominent at night during clear weather with low winds. While urban heat islands are often considered to be a problem during summer months, it is actually often during the winter when the effect is most pronounced.

Urban heat islands are often evaluated in terms of heat island intensity, which describes the maximum temperature difference between the urban and outlying regions. Urban heat islands typically exhibit a positive relationship with population size [1]. In Japan, most major cities are located in coastal areas, and there are many cases where urban development extends to mountainous regions with different topographical conditions. Heat island intensity in Japan is often generalized based on the quality of temperature contours, but it is often difficult to accurately specify an outlying, non-urban location.

12.1.2 Annual temperature trends

Figure 12.2 indicates annual temperature trends of several major world cities. The temperature in Tokyo (Otemachi district) has increased 2.9 degrees Celsius over the past 100 years. During this time, global temperature is estimated to have increased 0.7 degrees, which suggests that global warming in Tokyo is occurring at a speed of several times greater than that of the planet.

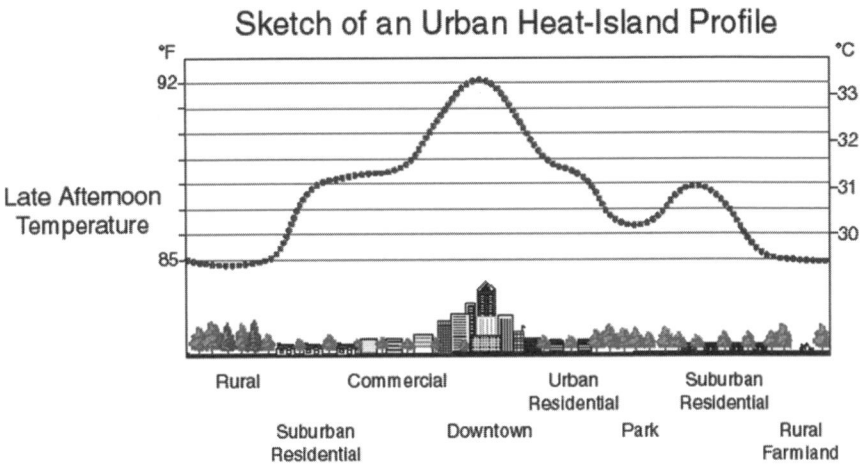

Fig.12.1. An urban heat island (Source: Lawrence Berkeley National Laboratory)

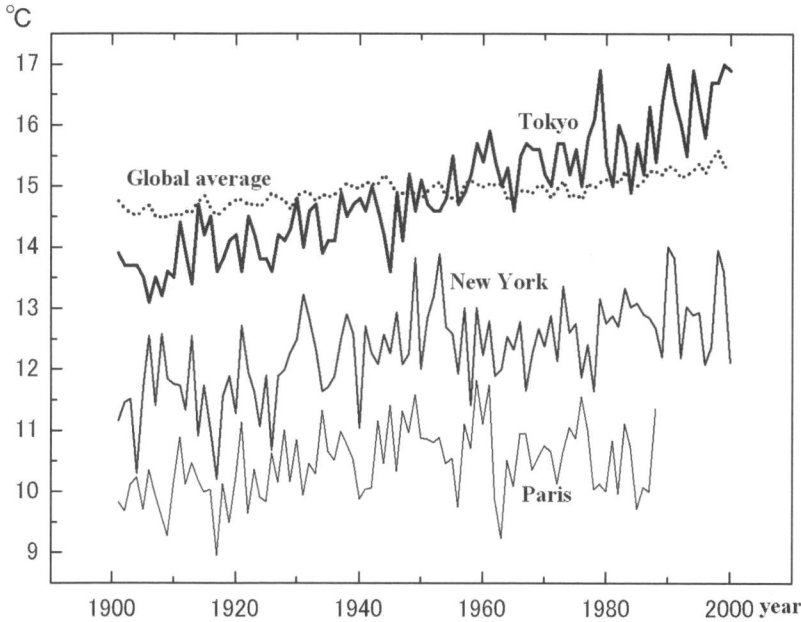

Fig.12.2. Yearly average temperature for Tokyo and other world cities (Original drawing: Mikami, Partially edited)

While temperature change in New York and Paris is not exceptionally noticeable after 1950, in the case of Tokyo the temperature has been continuously rising into the latter half of the 20th century. The reason for this difference is believed to stem from disparities in urban form and architecture found in Japanese and Western metropolises, as well as variations in the time periods of urban development.

12.1.3 Temperature distribution and winds in Tokyo

The distribution of summer temperatures and winds around Tokyo and the surrounding region is indicated in Figure 12.3[2]. The daytime peak temperature does not appear in the center of the city, but form a high-temperature region extending from the northern part of Tokyo's 23-ward area to Saitama prefecture located further inland. In the 23-ward area, a southerly wind predominantly blows from Tokyo Bay. Due to a difference in temperature between the ocean wind and the temperature on land of a few degrees Celsius, the cooling effect of the ocean wind is believed to be influencing this temperature distribution. Naturally, the cooling effect of the ocean wind does not take effect when weather conditions are such that a land-based wind blows from the opposite direction. When those conditions persist and Foehn winds descend on the city, temperatures in the city can rise to extremely severe levels.

During night hours, peak temperatures appear around the 23-ward area of Tokyo, creating a temperature distribution pattern that is completely different when compared with midday. Night temperatures differ from two to three degrees Celsius between the city center and outlying regions, which places almost the entire 23-ward area of Tokyo in an air temperature region marked by tropical nights, where the temperature is above 25 degrees Celsius. The heat island effect also causes winds to blow in from surrounding areas and converge on the city. Moreover, winds tend to blow in a counterclockwise direction, which is believed to be attributable to the Coriolis effect caused by the Earth's rotation.

12.1.4 Factors behind increasing urban temperatures

Figure 12.4 provides a diagram of factors that contribute to temperature increases in urban spaces. In addition to rising temperatures of ground coverings in urban regions caused by a transition to non-pervious coverings, anthropogenic heat emissions are also generated by urban activity. The amount of heat warming the atmosphere also exceeds natural limits. Further, building clusters decrease winds, which cause the ventilation potential of urban regions to drop. Combined with the greenhouse effect caused by fine dust particles, the resulting effect is an urban space prone to temperature increases. The individual factors contributing to higher urban temperatures are explained individually in the following sections.

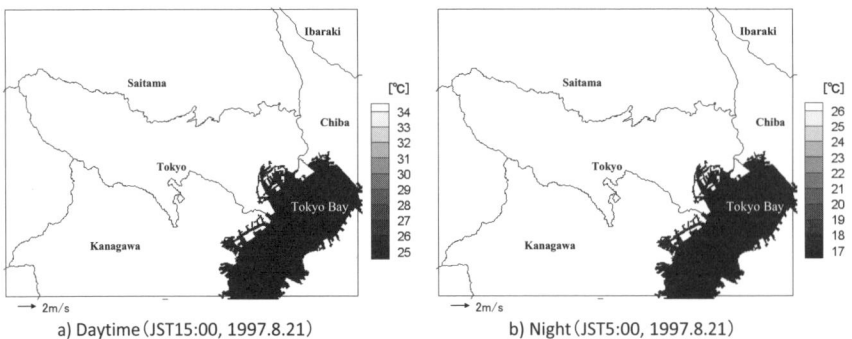

a) Daytime (JST15:00, 1997.8.21)   b) Night (JST5:00, 1997.8.21)

Fig.12.3. Examples of air temperature distribution and wind patterns based on ground observation (21 August, 1997). [2]

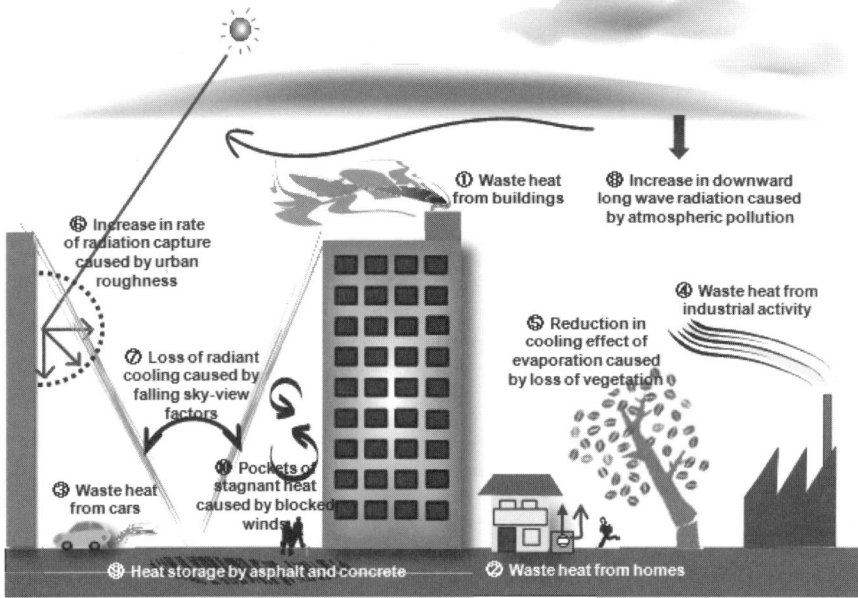

Fig.12.4. Factors behind rising urban temperatures

Ground cover

Figure 12.5 indicates the breakdown of land usage within the 23 wards of Tokyo. With 56.6% for buildings and 21.1% for road purposes, buildings and roads account for 77.7% of the total. Since buildings and roads primarily consist of man-made materials like asphalt and concrete, the surface temperatures of these materials when exposed to sunlight in the middle of the day can rise up to 50 or 60 degrees Celsius. The added thermal mass of the houses and roads means that even at night, the urban surfaces still fail to cool significantly. Combining the land use ratios of forests, water, agriculture, bare wilderness, and parks, the figure only amounts to around 13.8% for the whole 23 wards of Tokyo. Looking at the overall pattern of land usage in Tokyo, there is little chance for vaporization from such areas to cool the ground and other surfaces.

Anthropogenic heat emissions

The daily total of anthropogenic heat emissions (for an average day in August) generated from buildings, transportation systems and industrial facilities in all 23 Tokyo wards amounts to 2,000 TJ per day. When compared with the total amount of solar insolation received by Tokyo in August, this figure amounts to about 10% of the insolation on a clear day, or 20% of that on an average day in August. Consequently, the amount of heat released from human activity in Tokyo constitutes a heat source comparable with the amount of solar radiation received by the city, which is a fact that cannot be ignored. Furthermore, the contribution from buildings to the total of human-induced heat emissions is significant, contributing about half of the total, which makes buildings major culprits of heat island formation. Figure 12.6 indicates the distribution of anthropogenic heat emissions in the 23 wards of Tokyo [3]. The figure shows the daily totals on an average summer day of the total heat released to the atmosphere and in wastewater from buildings, transportation systems and industrial facilities. Anthropogenic heat emissions tend to concentrate in urban districts with high-rise buildings like Shinjuku, where the total is equivalent to the total amount of global solar radiation received by the same area.

Wind flow

The shape and roughness of urban surfaces is an important element in terms of considering wind flow. For the 23 wards of Tokyo, buildings occupy nearly 30 percent of the ground cover. While building heights differ for various reasons, buildings in Tokyo are three stories tall on average. Consequently, urban structures combined with the shape of local topography contribute to surface roughness. Figure 12.7 indicates the results of visualizing wind currents around a model building using a smoke in a wind tunnel test [4]. The test demonstrates an area of depressed wind speeds in the region downstream from the building. When similar weak-wind regions multiply in urban spaces, it is believed that heat released from surface coverings and air conditioning units gets easily trapped within the urban space.

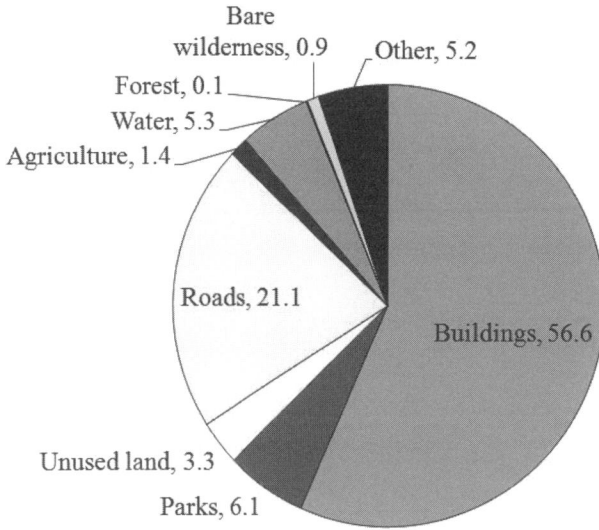

Fig.12.5. Land use allocation in the 23 Tokyo wards (%)

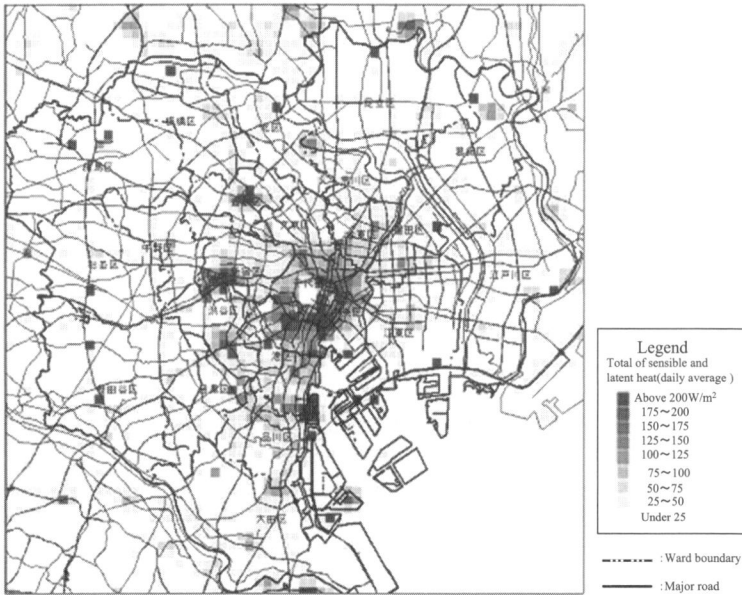

Fig.12.6. Anthropogenic heat emission in summer (Ministry of Land, Infrastructure and Transport. Ministry of Environment) [3]

Fig.12.7. Current visualization in a wind tunnel test (wind blowing from right to left) [4]

12.1.5 Implications of the urban heat island effect

Air temperature is a crucial environmental factor when it comes to humans and buildings. Since urban heat islands occur due to humans gathering in urban regions, the effects have direct and significant consequences for human beings. Furthermore, since the amount of temperature change clearly exceeds natural limitations, the environmental implications of urban heat islands are also of pressing concern. Table 12.1 provides a comparison of the summer versus winter environmental affects caused by urban heat islands to humans and the natural environment. Rising temperatures in the summer tend to generate scenarios with mostly negative impacts, such as rising demand for space cooling and more cases of heat related illness. In the winter, urban heat islands can have beneficial effects for humans like reductions in thermal stress, but there are also negative effects such as disease vectors surviving through winter periods. It is important to consider all of these environmental implications in their entirety. Secondly, it is also necessary to fully consider regional differences in urban heat island formation. Explanations of the main environmental implications of urban heat islands are provided below.

Implications for urban energy demand

Higher temperatures in cities during the summer translate into higher space-cooling loads for buildings. In the Southern Kanto region (where Tokyo is located), electricity demand increases by a total of 1.6 million kW for every 1 degree Celsius increase in temperature [6], which corresponds to the power capacity of two medium sized nuclear reactors. Figure 12.8 shows an example from Osaka of the temperature sensitivity of power demand. Here, the temperature sensitivity during summer is readily apparent. During winter, the demand for power rises with a fall in temperature, but it is not as pronounced as the change in power demand during summer. This example was limited to electric power, though sales volumes of gas and heating oil used for hot water and space heating tend to increase during the winter. Increasingly higher proportions of the power supply relying on electricity during peak cooling loads in summer also tend to reduce the annual operating capacity of power plants for other purposes.

Table 12.1 Environmental effects of the urban heat island effect (Ministry of Environment) [5]

Effect Item	Effect classification	
	Summer	Winter
1. Health effects 1-1 Health effects caused by changes in heat stress	* Increase in heat stress for those working in outdoor environments (outdoor labor, sports) * Increase in illnesses outside of directly heat related illnesses (e.g. circulatory system illnesses)	* Reduction in thermal stress for children, the elderly, and the economically disadvantaged * Reduction in disease from warmer winters
1-2 Health effects from increased usage of air-conditioning	* Physiological effects from air-conditioned spaces * Increase in heat stress from moving between air-conditioned space and outside	* Reduction in heat shock from moving between air-heated space and non-heated space.
1-3 Health effects from viral infections	* Rising fertility of disease-spreading organisms (increase in fecundity) * Viruses activated by climate change (rising air temperatures, low humidity) * Bacterial growth * Faster spoiling of foodstuffs	* Increase in the habitat and life cycles for disease-carrying organisms (such as mosquitoes) * Viruses activated by climate change (rising air temperatures, low humidity) * Disease-carrying organisms surviving in subterranean spaces.
1-4 Other health affects	* Sleep disorders	* Increase in respiratory illnesses caused by lower humidity.

2. Ecological effects 2-1 Change in habitat (weather conditions)	* Reduction in plant growth caused by hot conditions with low humidity * Ecosystem shifting north. Rise and fall of certain species. * Effects on waterborne organisms along urban coasts.	* Damage to dormancy cycles, dormancy disruption due to shrinking temperature range. * Ecosystem shifting north. Rise and fall of certain species. * Effects on waterborne organisms along urban coasts.
2-2 Change in habitat (mutual interactions among organisms)	* Changes to incubation, growth periods, and prey-and-predator relationships.	* Changes to incubation, growth periods, and prey-and-predator relationships.
3. Effects on the climate and atmosphere	* Localized sudden downpours caused by heat thunderstorms * Formation of photochemical oxidants	* Rising concentrations of atmospheric pollution in the mixing layer. * Changes to snow pack levels, duration, melt periods.
4. Effects on energy consumption	* Rising CO_2 emissions caused by higher demands for residential cooling. * Increase in the demand for water.	* Falling CO_2 emissions by lower demands for residential heating. * Falling CO_2 emissions by less energy consumed for water heating.

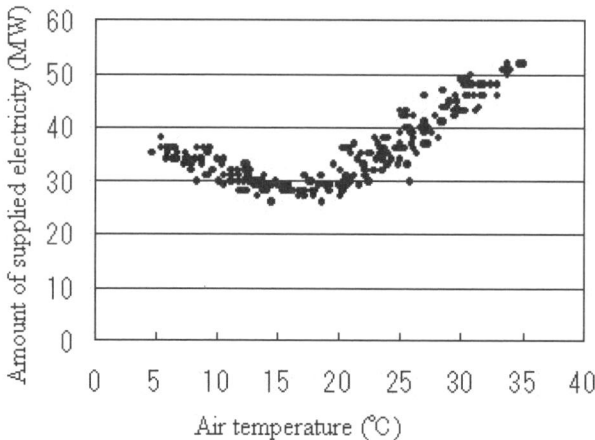

Fig.12.8. Relationship between electricity demand in office districts and air temperature (Original drawing: Narumi, Partially edited)
* From Kansai Electric Power data for 2003 (using data from 17:00)

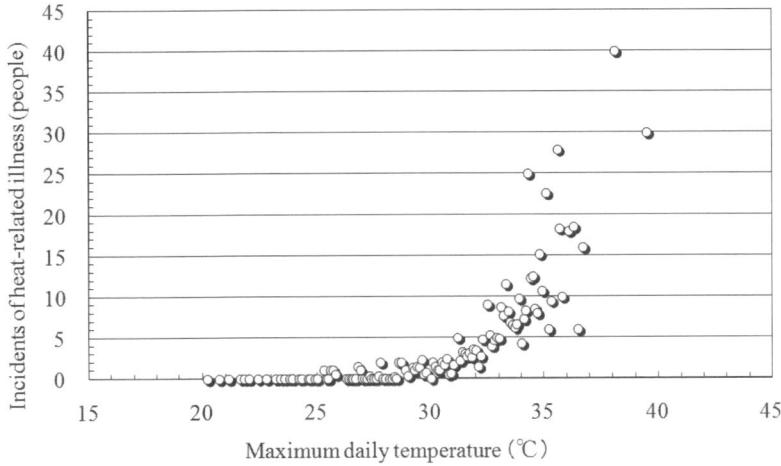

Fig.12.9. Relationship between numbers of people treated for heat-related illness and maximum daily temperatures in the 23 Tokyo wards. (Original drawing: Ono, Partially edited)
From records of ambulance calls submitted from the Tokyo Fire Department to the National Institute for Environmental Studies

Effects on humans

Higher midday temperatures during summer months and increasing numbers of tropical nights have implications for human lives such as uncomfortably hot conditions for urban living, as well as health risks like heat-related illnesses. According to documents from the Tokyo Fire Department, the number of people treated for heat-related illnesses doubled from 1980 to 2000 [7]. Figure 12.9 indicates the relationship between maximum daily temperature and the numbers of people treated for heat-related illness. Heat-related illness tends to occur after the temperature reaches 30 degrees Celsius, and continues to increase at higher temperatures. Diseases of heat-related illnesses, dehydration, and ureterolithiasis occurring during warm periods are indicated in association with physiological response to hotter temperatures such as increased perspiration.

Ecosystems

There are mounting concerns that changes in the temperature environment of urban regions are affecting both land-based and water-based ecosystems. Figure 12.10 indicates the yearly change to the northern limit of the habitat for the Asian tiger mosquito (*Aedes albopictus*) [8, 9]. Dengue fever is caused by humans becoming infected by bites from the yellow fever mosquito (*Aedes aegypti*) or the Asian tiger mosquito. While Dengue fever does not currently exist in Japan, an outbreak of Dengue fever did occur in 1940 caused by infection from Asian tiger mosquitoes. The presence of the yellow fever mosquito has also not been confirmed in Japan since the 1970s. This mosquito finds it easy to adapt to human environments since they can easily live inside cans or other small containers placed indoors, and it is said that eradication efforts in Singapore have been struggling. As heat island formation has pushed up the annual minimum temperature in Japanese metropolitan areas, there has been a four degree Celsius increase in annual minimum temperature in Tokyo during last century. For disease causing vectors, it is the annual minimum temperature that serves as a threshold for survival through the winter months. As heating trends continue apace during winter in major cities, the ability of disease causing vectors to survive through winter and establish further footholds is of pressing concern.

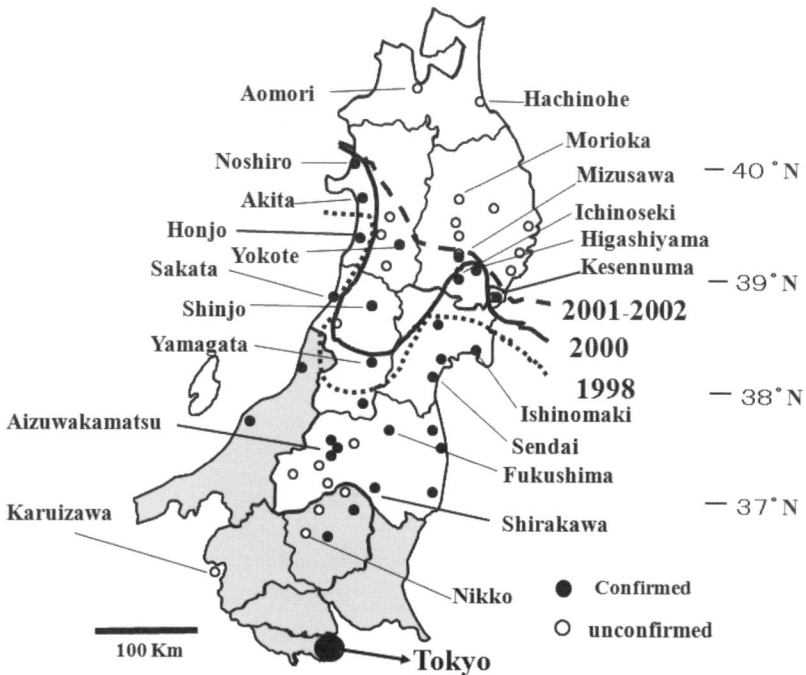

Fig.12.10. Shifts in the northern limit of Asian tiger mosquito distributions (Original drawing: Kobayashi, Partially edited) [8, 9]

12.2 Countermeasures and their effects on urban heat islands

12.2.1 Urban environmental planning and heat island counter-measures

Figure 12.11 lists urban heat island countermeasures according to three different scales, which can be represented by national environmental plans,

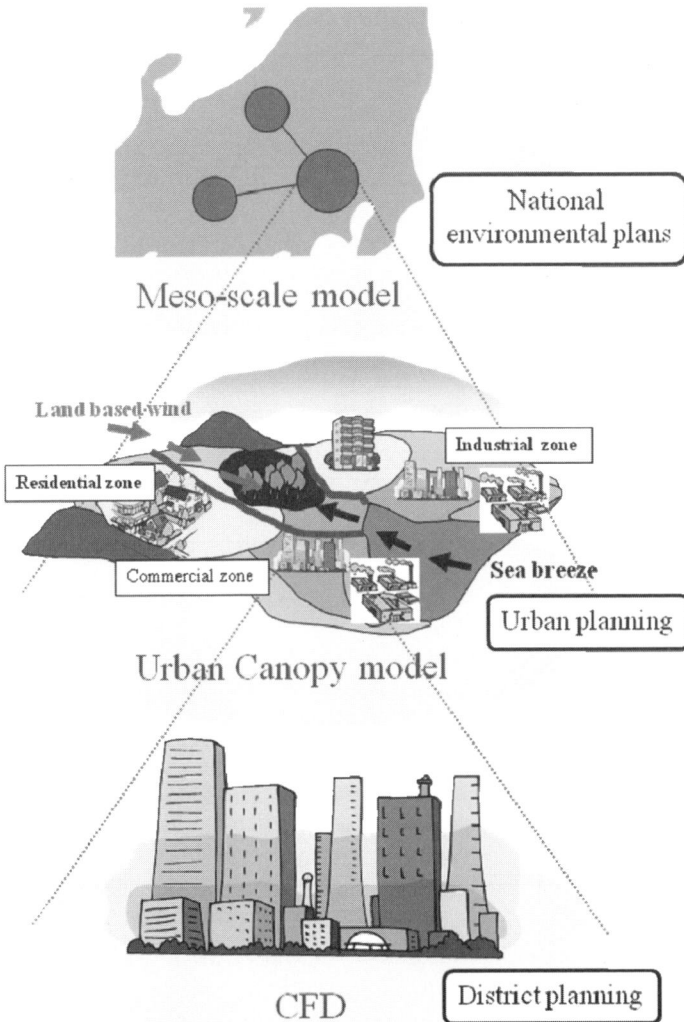

Fig.12.11. Numerical simulations methods at various scales

city environmental plans, and district environmental plans. Each plan addresses different elements, which differ according to spatial scale. Consequently, it is important to select an appropriate numerical model adequate for assessing a particular heat island countermeasure. Numerical simulations provide effective methods for evaluating the effects of heat island countermeasures and clarifying the urban heat island phenomenon. Typical numerical models include the meso-scale model, the urban canopy model and CFD (Computational Fluid Dynamics) assessments. Table 12.2 provides an explanation of how each model addresses aspects such as the definition of the urban structure, analytical resolution, region of analysis and primary input conditions.

Table 12.2 Numerical models used in heat island analysis

Model	Conceptualization of urban structure	Resolution	Region of analysis	Primary input conditions
Meso-scale model	City treated as a continuous surface with roughness	Rough	* Horizontal: Several 100 kilometers in all directions * Vertical: Several thousand to tens of thousands of meters	Physical properties (like heat conductivity, evaporation rates, roughness, and albedo levels) and anthropogenic heat emissions input for different land use types (built up areas, agricultural land, oceans and mountains)
Urban canopy model	Building clusters represented with average densities and heights	Middle	* Horizontal: Several tens of kilometers in all directions * Vertical: Several hundreds to thousands of meters	Building cluster conditions (building-to-land ratios, building height, AC systems) and physical properties of land cover (heat conductivity, evaporation rates, albedo levels) input
CFD	Building shape and placement are represented	Fine	* Horizontal: Several hundreds of meters in all directions * Vertical: Several tens to several hundreds of meters	Shape and location (of buildings, roads, trees, etc.) as well as inputs for physical properties of individual coverings (heat conductivity, evaporation rates, roughness, albedo levels, etc.) and anthropogenic heat emissions

Generally, national environmental plans treat cities as the subject of the plan and investigate environmental problems over an expansive area associated with city placement, capital relocation or flows between cities. Meso-scale models are suited for analysis at the national environmental plan level, and provide justification for environmental planning at the national level by investigating the conditions of sea and land-based winds based on geographic location.

City environmental plans focus on land use zoning, and are used to analyze spatial relationships between land use zones in order to facilitate regional planning of road networks and green belts. Methods for generating winds by strategically placing river ways and open space such as large parks, or spatial relationships of inner-city air pollution due to locations of industrial facilities and road systems are investigated at the city environmental planning scale, which are sometimes used to justify environmental improvements using restrictions to floor-to-area ratios and building heights. In such cases, meso-scale models can be used to provide boundary conditions, and then numerical simulations from urban canopy models can be used as a supporting technology for urban environmental planning to assess various heat island countermeasures. Such models can also be used to provide recommendations for detailed district planning. These types of planning policies and examples of urban meteorological information can then be organized into a series of maps known as urban environmental climate maps.

District environmental planning occurs at a more realistic scale in terms of urban redevelopment, and provides detailed planning concerning building cluster alignment, or the placement of parks and trees. The main constituents of district level planning are urban developers and local community organizations, and concern a wide range of stakeholders and urban design initiatives, from local citizens to NPOs. CFD is used for most numerical analysis at the district level of environmental planning, because it can divide the space between buildings into multiple meshes to perform detailed analysis of the flow of wind and heat around buildings.

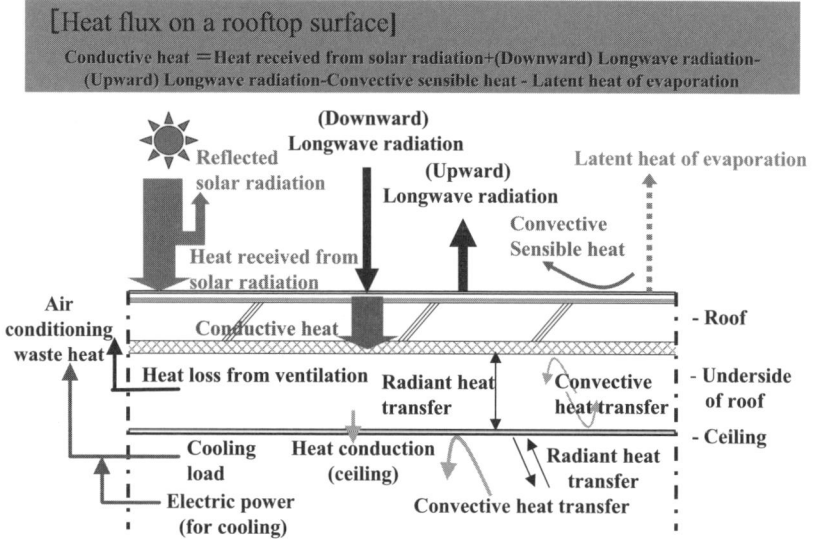

Fig.12.12. Heat flows at a building rooftop at midday

12.2.2 Thermal composition of building rooftops

Figure 12.12 indicates thermal flows at a building rooftop. Of the solar radiation received by the roof, some will escape into the atmosphere by convection and evaporation and some will be released as long wave radiation. The remaining portion will enter the building as conductive heat. The indoor cooling load of a building works by powering air-conditioning equipment with electricity to expel the internal heat to the outdoors. Consequently, engineering the heat flux at the rooftop surface to enhance evaporation or reflection and reduce the amount of thermal load received from radiation can be an effective way to reduce heat dissipated to the building from the rooftop. Additionally, providing insulation to lower thermal transmission at the roof can reduce indoor cooling loads and, in turn, reduce waste heat from air-conditioning systems.

12.2.3 Rooftop gardens

As a countermeasure to the urban heat island, the Tokyo Metropolitan Government now requires at least 20% of the surface area of accessible rooftop space to be garden space for all new buildings on sites that are 1000 m^2 or larger. As a result, demand is growing for rooftop garden systems that are low cost and require little maintenance. Succulent varieties of plants such as those in the *Sedum* genus are typically resilient to dry conditions, and most will survive for at least one year even when planted in lightweight soils. The Japan Urban Renaissance Agency has independently developed technologies for maintaining grass in a shallow soil substrate, and is now presenting them as standard specifications for rooftop garden systems. An example of a shallow bed rooftop garden system is depicted in Figure 12.13. Unfortunately, sometimes management issues at multi-unit housing complexes of the Urban Renaissance Agency prevent people from accessing the rooftops, which prevent the buildings from being counted as rooftop space for rooftop gardens by the Tokyo Metropolitan Government. As of 2004, green space (counted as the portion of surface area covered by vegetation and water) in the 23 Tokyo wards was 29%, but Tokyo aims to raise this number to 32% by 2015. Doing so will require installing 1,200 ha of rooftop gardens in Tokyo, a major increase from the current figure of around 23.7 ha (MLIT, 2005).

12.2.4 Cool roofs

Cool roofs typically refer to rooftops that reflect higher ratios of solar radiation. By reflecting solar beams, cool roofs suppress temperature increases of rooftops, which it is hoped will help cool urban temperatures and reduce cooling loads. In the United States, the Environmental Protection Agency is promoting both tree planting and the use of light-colored shades on building surfaces to reduce the urban heat island effect. Paint that increases the amount of reflected radiation can be referred to either as "highly reflective" paint or "heat-shield" paint. Solar radiation that reaches the ground is composed of electromagnetic waves of varying wavelengths, with most of that radiation composed of equal portions of visible light (400 – 760 nm spectrum) and near infrared radiation (760 – 3000 nm spectrum). By using special pigments, it is possible to create paint with wavelength attenuation that reflects more light in the near infrared portion of the spectrum that is invisible to the human eye. Figure 12.14 demonstrates two paints with similar spectroscopic properties in the visible portion, but one paint uses wavelength attenuation to reflect nearly two times as much radiation despite having the same visible color properties.

12.2.5 Building energy flow

Understanding flows of energy and heat within a building requires looking at these concepts in terms of three vantage points: the supply phase, the consumption phase and the emissions phase. The supply phase can be assessed in terms of CO_2, the consumption phase in terms of energy and cost conservation, and the emissions phase in terms of the urban heat island effect.

Fig.12.13. Rooftop garden (photograph from the Urban Renaissance Agency)

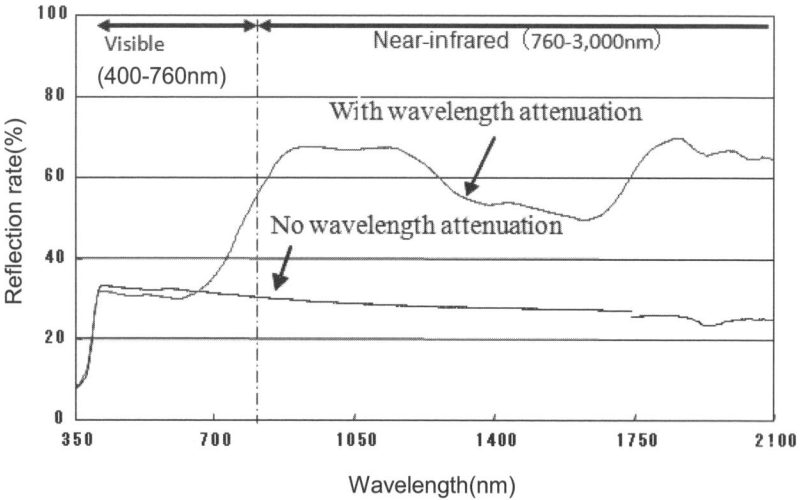

Fig.12.14. Paint wavelength attenuation (Original drawing: Matsuo. Partially edited)

Total heat kWh/Day			kWh/Day			
Building received	(INPUT)		From building facilities	(OUTPUT)		
			Sensible heat	Latent heat	Hot waste water	Total
Solar radiation	19,576					
Ventilation	2,092					
Electricity	36,775		51,140	7,414	546	59,100
Fuel	658					
Total	59,100					

Atmospher

Loss from power transmission 2.5%
【Tokyo Power】
Transmission end efficiency: 39.12%
【Federation of Electric Power Companies】

A : Solar radiation 8,659
B : Heat transfer 3,877
C : Body heat 7,039
14,972
E : Motive energy (AC) 4,035
G : Lighting 12,228
H : Power outlets 5,540
I : Electric kitchens 0
J : Gas kitchens 0
K : Boiler 546 112 0
L : Hot water supply 546

100,423

Power plant 36,775 Electric power

Petroleum Natural gas

658 658 658 546

Food

(A+B+C+D+E+G+H+I+J)
43,470

Air-conditioning load

Building

D : Ventilation -1,287 -805
2,092

(AC=Air-conditioning)
F : Motive power (Non AC) 1,729
11,681

M : Package AC 46,803
35,122
P : Electric chilling unit (including heat storage tanks) 1,563 1,172
8,348
8,379
N : Air cooled outdoor cooling unit 0
Q : Steam / hot water absorption refrigeration units 0 0
O : Cooling tower 1,376 7,363
Gas emissions
R : Direct heating absorption refrigeration unit 0
0 60 51 112

Atmosphere
sensible heat   latent heat

Ocean

Sewage

Petroleum Natural gas

Electric power

Heat

Primary energy (supply phase)
【CO2 assessment】

Secondary energy (consumption phase)
【Energy and cost conservation assessments】

Tertiary energy (emissions phase)
【Urban heat island assessments】

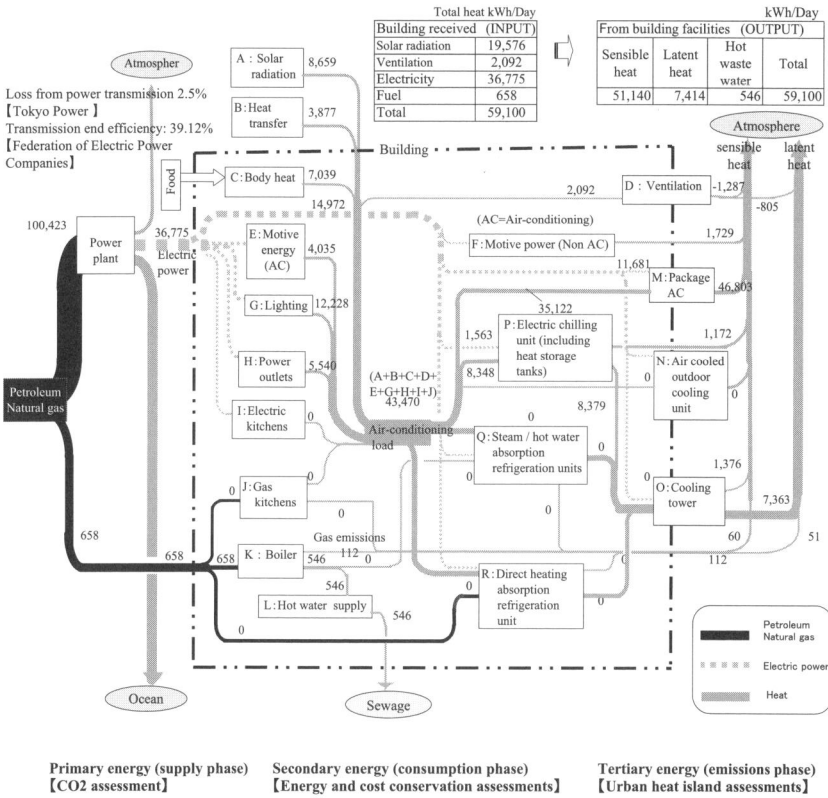

Fig.12.15. Building energy flows (including waste heat) [10]. (Example from an actual building in Tokyo)

Figure 12.15 contains an example of the actual energy flows (including waste heat) of an actual building in Tokyo explained in terms of the air-conditioning system in the supply, consumption and emissions phases [10]. The amount of energy used in the consumption phase can be examined to assess the net increases in consumption of electricity and gas, but this fails to reveal the amount of sensible and latent heat emitted by building facilities. The amount of sensible and latent heat released from building facilities can be calculated by considering the heat dispersion characteristics from a building's cooling towers or outdoor cooling units.

The amount of heat entering the building and waste heat released from the building equipment are equal.

Heat that enters a building   (1)

= Heat from the consumption of electricity and gas

+ heat transmitted to the building from radiation through windows, and absorbed by the building through ventilation

= amount of heat released from the building equipment

Since the heat a building absorbs from the surrounding environment is released as latent heat and not dispersed as sensible heat, building equipment operates on a similar principle as rooftop gardens. The amount of heat released from building equipment should be used in order to assess the performance of handling heat absorbed in such an environment.

12.2.6 Urban wind-passage and the heat island effect

The first skyscrapers appeared in the Kasumigaseki area of Tokyo in the 1960s. With them came the introduction of "building winds" as a social problem. "Building winds" is a phenomenon in which extremely strong winds confined to a small, localized area are generated due to wind striking the face of a building, blowing downwards, and then wrapping around the sides of the building. As problems from building winds started to become more apparent, people started to actively seek solutions to address the problem by engineering the shapes of buildings or planting trees. Currently, environmental assessments into the potential for wind-related hazards are performed when constructing buildings exceeding a certain size.

In recent years, urban redevelopment initiatives have sparked a resurgence of skyscraper construction projects in urban cores. The Shiodome district built along the Tokyo Bay represents a typical development project, and is represented in an aerial view using digital data in Figure 12.16. As construction of groups of skyscrapers continues, it is believed that higher building densities will lead to lower wind speeds within the city. Consequently, cities are entering an era in which the on-going construction of skyscrapers is exacerbating the increase in urban temperatures, and where urban structures are believed to be behind problems associated with weaker wind environments. The creation of wind passage which can introduce sea breezes inside urban spaces must be an effective way to reduce urban heat island in Tokyo Bay area.

12.2.7 Evaluating urban environments with numerical simulations

Recently, vector-parallel supercomputers such as the Earth Simulator have been used to perform numerical simulations at the global scale, giving rise to a flurry of new research activity of various physical phenomena. Leveraging this type of large-scale analytical capacity to study engineering issues at the scale of urban construction allows for the tackling of several environmental problems that have hitherto remained beyond the processing limitations of conventional computing resources.

The cooling effect of ocean winds blowing into urban districts on summer days is being looked at as a way to mitigate the urban heat island effect. Urban areas contain enormous amounts of buildings, along with rivers, parks and open space. These spaces were represented with fine-scale meshes in a massive CFD analysis, that was performed using the Earth Simulator under the following conditions.

- Analysis region: 6 km (X: East - West) × 6 km (Y: North - South) × 500 m (Z: Vertical)
- Computational grid size: 1,200 (X) × 1,200 (Y) × 100 (Z)
- Mesh cell width: Horizontal equal interval of 5 m, Vertical unequal interval of 1 m ∼ 10 m
- Input wind: power law, neutral conditions
- Wind speed ground boundary: Generalized log law
- Heat ground boundary: Temperature fixed for each land use, heat transfer boundary
- 20 nodes used (160 processors)

Fig.12.16. Aerial view of Shidome area (using CAD CENTER digital data)

28 29 30 31 32 33 34 35 ℃

Fig.12.17. Example of urban temperature analysis using Earth Simulator [11]

Figure 12.17 indicates the horizontal distribution of temperature at a height of 9.7 meters [11]. The wind direction is blowing as a southern wind, from the bottom of the image towards the top. Temperatures along the coast and above waterways appear clearly lower. This image demonstrates an example of analyzing urban temperatures where actual building positions are regenerated in detail, which would not be possible using a meso-scale model.

References

[1] Landsberg HE (1981) The urban climate. Academic press
[2] M. Moriyama: "Heat island countermeasures and technologies" (*Hi-to airando taisaku to gijutsu*), Gakugei Shuppan Sha Co., Ltd. 2004.8
[3] Ministry of Land, Infrastructure and Transport / Ministry of the Environment: "2003 Investigative report into restricting anthropogenic heat emissions as a countermeasure to the urban heat island effect in cities" (*Heisei 15 nendo toshi ni okeru jinko-hainetsu yokusei ni yoru hi-to airando taisaku chosa hokokusho*), 2004.3
[4] Y. Ashie, S. Yoon: Examination on the influence of the development of a high-rise building on air temperature distribution by means of floor-heated wind tunnel test (*Yukamen wo kanetsu shita fudojikken ni yoru tatemono no kosoka ga kion bunpu ni oyobosu eikyo ni kan suru kento*), Journal of Environmental Engineering, Architectural Institute of Japan, No. 579, 2004.5, 67-71
[5] Investigative committee on the environmental effects of urban heat islands · Center for Environmental Information Science: 2003 Activity report on investigations into the environmental effects of the urban heat island effect, 2004.3
[6] Ministry of the Environment: 2000 Report on state of analysis and countermeasures of the urban heat island effect (expanded edition), 2001.10
[7] Cited in [6] above
[8] Infectious Agents Surveillance Report, Vol.25, No.2 (No. 288) Infectious Disease Surveillance Center, 2004.2
[9] M. Kobayashi, N. Nihei and T. Kurihara: Analysis of northern distribution of *Aedes albopictus* (*Diptera Culicidae*) in Japan by geographical information system. Journal Medical Entomology, 39(1), 2002, 4-11
[10] Y. ASHIE, M. TANAKA, T. YAMAMOTO: Order analysis on the exhaustion of building equipment considering the origin of natural and machinery system (Research on the thermal metabolism of large scale buildings Part 1) (*Shizenkei oyobi kikikei no yurai wo koryo shita kenchikusetsubi no hainetsu no oda bunseki – daikibo tatemono no netsutaisha tokusei ni kansuru kenkyu Sono 1*), Technical Journal of The Society of Heating, Air-Conditioning and Sanitary Engineers of Japan, 2004.9, 1059-1062
[11] T. KONO, Y. ASHIE, S. YOON, H. LI: CFD analysis of airflow and temperature at the urban scale by a 5 m resolution (*Toshi-sukeru wo taisho toshita 5 m messhu kaizodo ni yoru fusoku / kionba no CFD kaiseki*), 18th Symposium for the Japan Association of Wind Engineering, 2004.12, 117-120.

Index